A Perthshire Beekeeper

Athole Kirkwood's beekeeping diary

Edited by Fiona Mackenzie

A Perthshire Beekeeper – Athole Kirkwood's Beekeeping Diary

Obtainable from: www.northernbeebooks.co.uk © 2026

First published in 2026 by
Northern Bee Books
Scout Bottom Lane, Scout Rd, Mytholmroyd,
Hebden Bridge HX7 5JS
Tel: 01422 882751

ISBN: 978-1-918815-01-6

Edited by Fiona Mackenzie
Designed by Ian Kirkwood www.ikdesign.scot

Front cover photo 1971, *Dundee Courier*
"Speaking Personally by Fiona"
by kind permission of D.C. Thomson & Co Ltd.

For Barbara

A Perthshire Beekeeper

Athole Kirkwood's beekeeping diary

Contents

FOREWORD

70 Years with the Bees!

"Write a page a day!"

This was the rather ambitious note I found on my father's desk when he passed away at the end of March 2020 at the age of 92. His family and friends had often encouraged him to put down in writing his own account of his fascinating beekeeping career and although he may have had great intentions of perhaps one day writing a worthy companion to the many well-thumbed books in his library, including those by various famous names in the beekeeping world, particularly his idol R.O.B. Manley, and numerous others, when it came down to it, writing about his life seems not to have been nearly as entertaining as actually living it. His fondness for an exclamation mark was evidence of his enthusiastic approach to most things in life (fishing, photography, antique hunting (including his fascinating collection of Glasgow Empire Exhibition memorabilia), mathematics and chemistry, keeping an eye on the stock markets, art, a good long walk and of course his family) but perhaps writing lengthy detailed memoirs was not one of them. He did in fact begin a collaboration on a book with his great friend, Francis Cordingley, but it seems that project was doomed to gather dust many years ago.

1962 © Dundee Evening Telegraph

Perhaps he did intend to persevere with his story at some point, possibly imagined there would be more time, (we thought so too) but his first pages on his childhood were laid in his "to do" pile several years ago. He had actually completed almost 75 years of beekeeping by the time he finally "laid aside his smoker", and, as we discovered, he hadn't really got around to expanding on the nub of the tale – the bees!

So now we are left with the puzzle of how best to complete the book he had really only just begun, a tale we all fervently believe should be told, but how did he plan to enlarge on his unusual story? The only clue we have as to how he intended to proceed is that his note included an instruction that "it should be written as a diary". We have the years from 1945 until 1956 in diary form, at which time he married our mother and began to build up the business in earnest, but after that, his annual notes are unsurprisingly more brief.

His beekeeping techniques were built on the sound foundations of his College course, and modified with advice and suggestions from many knowledgeable and experienced practitioners of the craft but were largely his own. His son is himself a beekeeper and learned all he knows from our father, having worked alongside him for many years, sharing in his great store of beekeeping lore, but it was inevitable that our father, having lived to such a ripe old age, would sadly outlive nearly all of his contemporaries so we are unable to turn to many of them for their recollections.

Along with his enthusiasm for an exclamation mark, however, went an astounding talent (or perhaps obsession) for note making. We, as children, were all aware of the notes: reminders and jottings all about the house, the garage, the sheds, the honey farm, on the stairs and pinned in cupboards with instructions and dire warnings of every description. But I don't think any of us had realised until sorting through his belongings that he was quite such a hoarder. He had kept all of his diaries – daily diaries of family events and various holidays and trips, brochures, articles, newspaper clippings (including those in which he had appeared), his machinery and equipment purchases, correspondence with friends and relations, various collections of sundry beloved bits and pieces, all meticulously labelled and also, more relevant to this proposed book, his yearly honey production records from acquiring his first hive in 1945 until 2018 and correspondence from honey producers and equipment manufacturers from all over the world together with countless photographs and slides.

Given that the lion's share of the words in the following pages are by no means those of the present editor, and, in tribute to "Uncle" Bob Cousten, "The Beeman", (recognising that there are many, many other Beemen), the editor will refer to herself as "the bee-daughter" from this point forwards. Although she still possesses a full bee outfit and might be of some limited assistance in an apiary, most of her years involved in bee farming took place in the extracting room, and it is hoped that readers will forgive any inadvertent errors made in the descriptions of the management of the bees themselves.

We commence with the beginnings of the book our father had in mind, describing his early life in Glasgow and evacuation to Strathardle, followed in Chapter 2 by the detailed annual diary he kept when he came to realise that beekeeping was to be the life for him. This takes us up to 1956, during which time he learned a great deal from his time spent at Auchincruive West of Scotland Agricultural College, a stay with A W Gale in Wiltshire, correspondence with many other very experienced beekeepers, and, of course, much trial and error.

The second part of the book, although it contains his annual notes, is not a diary in the same sense as the first. This period, of course, signified an enormous change in his personal circumstances, now being an extremely busy married man with, eventually, five children, to provide for! This was the heyday in the growth of Heather Hills Honey Farm and the years during which he gained respect for his knowledge and practice of the craft.

This is not the book our father would have intended, we are sure, which we imagine would have included far more technical information and the methods he developed over the years regarding management of his hives. In his beekeeping, he was never attracted by complexity. "Simple things for simple people", he would say, laughing at himself. Obviously he made subtle changes and adaptations over time and the records we have may only have applied during a particular period. We have, however, gathered together a varied collection of diary entries, photographs, records and snippets (presumably intended for "the book") representing a long and hardworking life, full of gentleness and humour but with a determination to succeed which could never be thwarted. His brief list of notes for his proposed book culminated in the statement that "Determination" was most important of all – this was noted with a very determined exclamation mark!

Determination!

INTRODUCTION

Dad was born in Selkirk on the 3rd October 1927 to Major William Kirkwood MC and Elizabeth Murray. From there they moved for a time to George Square in Edinburgh and then on to Glasgow where he found a circle of boyhood friends (Eddie, Peter and Alfred) whose sole purpose in life was to create mischief. Safety conscious parents of today would tremble at his love of "chemistry", "to assist in the war effort" resulting in several minor explosions, and amusements such as rearranging the track of a friend's prized railway in order to gain much pleasure watching it hurtle through a first floor window.

The earliest Kirkwood family records stretch back, on his grandfather's side to coal mining in Lanarkshire, followed, for some generations, by a spirit merchants' business in the Gorbals. Then, though two of his great uncles emigrated to seek their fortune in Australia, his grandfather, also named William Kirkwood, moved to Blairgowrie to open an ironmongers' shop in Allan Street. In contrast, Dad's grandmother's family originated from the hills and moorlands around Blair Atholl and Strathardle and when his grandfather died at the age of 52, she moved back to Kirkmichael, the village of her birth, where she remarried.

Dad's forced removal from the city to his grandmother's home in a Scottish glen at the age of thirteen due to the outbreak of WW2 was a joy. Here one's time was spent with a split-cane fishing rod in hand exploring the sources of burns and discovering remote lochans, then fetching home brown trout for tea. He once mentioned that as a young man his only ambition had been to find a way to return to those idyllic times in Perthshire, even if he could

Athole's grandfather, William Kirkwood (a striking resemblance!)

only earn enough to live in a "butt and ben". One day, he would indeed return and build a home there.

Many years were to pass, however, before that dream could be realised. That clear vision and an unwavering ambition to succeed drove him through all the ups and downs, the giddy heights and gut-wrenching lows encountered in the precarious career he had embarked upon.

Through it all, he could not have succeeded without the support and strength of our mother, Barbara, of whom he was fond of saying, "I put her on a pedestal the night I met her and she has been on it ever since". He married the former Smedley's 'Friday Girl' on 27th April 1957 and after a few years of saving hard, bought Glenardle, a spacious Victorian house in Bridge of Cally, north of Blairgowrie. "I would not advise buying this property at all Mr Kirkwood", said the Bank Manager, "The feu duty is £30 a year!". With mum's help and artistic flair the house became a wonderful family home and we five children enjoyed a fortunate life of freedom to roam the hills and rivers of the glen.

Dad was a single-minded man of high energy and determination. He was quite shy and diffident, quiet, always polite and gentlemanly, but he could also be impatient, pedantic and perhaps overly meticulous, and sometimes, on a tough day, younger helpers might feel discouraged. He found it perplexing that not everyone was as fascinated by the bees as he was himself, but he was exceptionally generous with his time and knowledge if he felt that a person was genuinely interested.

Despite his thin, wiry build he was capable of perpetual and tireless manipulation of the hives, hive parts and heavy pails of honey. It is doubtful whether anyone who ever assisted him felt that they were called upon to work harder than he did himself.

William Kirkwood's ironmongery in Allan Street, Blairgowrie.

As children, we were blissfully unaware of the fragility of the honey bees and the unpredictability of the Scottish climate as we set off for our holidays, but that was only at Easter or in October, during the tattie holidays. Summer holidays were rare as hens' teeth and Dad could never accompany us. Work on the farm when the bees were taken out to the heather started early and finished late, wet or fine. At the beginning of July he, along with Bob Barrie, Harry Kean, Allan Grant and other permanent staff, together with Willie Henderson and the much needed student help who joined them during the busy season, would be up at 3.00am. He would try to be home by 6.00 in the evening but it was often nearer 9 o'clock.

In spite of his success and his optimism that each season would be better than the last, there were, of course, many stumbling blocks to be surmounted

Wild flowers at Whitefield Castle, Strathardle.

along the way, but these were largely overcome in a gritty and methodical manner and his outlook was always hopeful.

Although he maintained his unwavering upkeep of the business, he was able to switch off completely and immerse himself in family and village life. He could relax at the drop of a hat, falling asleep in a chair (or anywhere really) while we listened to him snore. It must be said, he deserved some rest. He would take great care preparing for any family outing and had jackets for every occasion – he had a liking for tweed and many of them were quite stylish. As a neighbour recalled recently, his appearance was somewhat that of an estate factor. His daily work uniform consisted of light trousers (rarely jeans) and a well worn jumper. He had a great fondness for a ragged blue anorak he wore for all the really messy jobs. One day a floating ember from one of his bonfires landed on the back of this garment. When he returned to the house he appeared to have received a blast in the back with a shotgun.

He was generally a plain eater, with porridge and oatcakes being typical favourites, however when he came in from the farm for lunch he enjoyed some peculiar foods the rest of us would not touch, such as pickled ginger and a hard orange Norwegian cheese with his *Spam*, washed down with *Woodpecker* cider.

He was extremely partial to ripe bananas, which were always part of his packed lunches and was known to secretly stash Cadbury's milk chocolate in his office for rationing himself one piece at a time – very occasionally, our ears would prick up as a rustling of paper meant that we were to be allowed a piece too.

Whenever he had time he would haunt the auction sales and was always delighted when he found a new painting or piece of furniture. He and mum had very definite ideas on their taste. His eye for detail and appreciation for art and antiques meant that he designed two houses around his prized dining room table! Antiques were their main purchases but they also both appreciated beautiful contemporary crafts and paintings so our home was filled with their interesting collections.

We would often be rewarded with his unique knowledge of the history and peoples of his beloved hills and glens; on the other hand, he greatly enjoyed travelling to foreign climes (but of course was not fond of the food – his mother's gooseberry tart could not be bettered)! He was always excited by birthdays, weddings, family events and visits to and by friends. He was full of chuckles and sometimes side-splitting giggles with numerous amusing stories and anecdotes.

But even on those occasions, when he had put the bees to the side for a time, one thing we know for sure, they were always on his mind – as he once shared with the bee-daughter, he absolutely "loved the bees".

CHAPTER 1

Early Years

As I recall, I first became aware of bees back in 1938 when, at 10 years old, I visited the Scottish Beekeepers' Association stand at the Empire Exhibition, Glasgow, though it made no great impression on me at the time. My father was the Exhibition Secretary, so all the family had season tickets. We went almost every day and greatly enjoyed all the stands. In particular, I did like the free samples of Canadian clover honey! And seeing the Canadian mounted police and their lovely horses.

My father, who had been a Major in the First World War, and who was awarded the Military Cross, became a Private in the Home Guard. He stayed on in Glasgow to run the Hillington Industrial Estate, much of which at that time was occupied by Rolls Royce, making and maintaining engines for the Hurricane and Spitfire fighter planes.

The Scottish Beekeepers stand at the Empire Exhibition, Glasgow.
Attributed to Dr Tennant, Member of Glasgow & District Beekeepers Association and President of the SBA.

The year following the Empire Exhibition brought the war, and in September my mother, myself and younger brothers, Harry, and Ronald (newly born), were evacuated to Perthshire, to stay with my grandmother for the next three years. My older brother, Ian, 18, was in the Army, my sister, Muriel, 17, in the WAAFs.

Kirkmichael was where my grandmother was born, and where she spent most of her old age. She was born a Campbell. Her father ran the Balvarran Estate and most of her relatives had worked in the area since at least 1780. Kirkmichael village lies at the heart of Strathardle, a beautiful glen and a marvellous place for a boy to grow up. The farms at that time were worked with horses, lovely great beasts that ploughed and tilled the land.

About fifty children from Glasgow were evacuated to Strathardle in 1939. Around half soon went home, but we stayed for the duration of the war. The younger ones went to the local schools at Kirkmichael and

William C Kirkwood during WWI.

Me at age 6 with Harry c. 1933.

Ballintuim, the older ones to Blairgowrie High School about thirteen miles away, where my favourite subjects were woodwork and metalwork, although I always thought I might have liked to have been a chemist. The French teacher, Miss Clark, a particularly efficient wielder of the tawse, turned out eventually to be my future wife's aunt. The bus was "utility" with hard wooden seats, but most days we sang all the way to school and all the way home, all jovial songs such as "We are going to hang out the washing on the Siegfried line". We never considered defeat!

The winters were long and hard. In 1940 or 41 there was so much snow that vehicles could not get to Blairgowrie. We were off school a month and spent our time mostly sledging or skating on the curling pond. A horse and carrie sledge was the only contact with Blairgowrie. It came up with bread, vital provisions and medicines. The snow was up to 6 foot deep in many parts. I spent a lot of time reading patiently through the 1927 edition of the Encyclopaedia Brittanica!

My younger brother, Harry and I shared a bed in a small room upstairs with a hinged window in the ceiling. One morning we woke up with a two foot drift of snow on top of us. The window had been open just a couple of inches!

The summers were wonderful, bursting with exciting activities such as trout fishing in the Ardle, trying to shoot rabbits with a cheap air rifle, helping to tread the hay on the big stacks and exploring the ruins of old

Pittarig, Pitlochry (between the wars).

With Grandmother, Ian and Muriel 1934.

Mum with my youngest brother, Ronald.

Mrs McLaren's bees, Burnside, Kirkmichael.

Whitefield Castle. We followed the Home Guard almost everywhere. They were a useful lot – keepers, gardeners, foresters and farm workers. With a popular rifle range at Enochdhu, just up the road they were mostly very good shots and ready for the Germans! Sometimes they used live ammunition in their rifles, mortar and one Lewis machine gun.

Whitefield Castle, Kirkmichael.

We often knew where they were to be going and hid not too far from the target. We had a slightly scary time when they attacked Whitefield Castle with live ammunition – we were watching from behind a dyke just a few hundred yards from the target! On another occasion I sat with the machine gunner when we ambushed a "German" convoy on the road – a great experience but with dummy cartridges!

During the war the Germans dropped thousands of bombs in the area – mostly incendiary, but perhaps 20 or 30 high explosive bombs. Most were dropped by planes after missing Clydebank. They were jettisoned over the hills so the planes could get home faster!

My younger brother, Harry and I used to go up the next day on to the moors to find the craters and collect shrapnel. Where the incendiaries hit the ground there would just be a burnt area about a metre in diameter and a metal tail fin sticking out of the ground. We saw hundreds of these when grouse beating in August/September.

A favourite pastime was to go down to the Ardle and spend happy hours fishing there. Once I walked right through a swarm of bees that had settled unseen on the river bank and they were inside my shirt before I could move. I ran as fast as I could back home and jumped in the bath. I received about 50 stings that day! Almost all the neighbours kept bees, and I recall attending a fascinating lecture and film show in Kirkmichael Hall on the subject, but I did not take any serious interest in beekeeping until the family moved back to a new home – a small rented estate at the beautiful Hill of Birches, Thornton Hall, just outside Glasgow where my father ran a select herd of about 300 pedigree Large White pigs.

There happened to be an empty beehive at the bottom of the garden and my father asked me to make some use of it. Although the house we moved into at that time was quite empty, by the telephone lay a copy of Websters 'Beekeeping'. From that moment on I became more and more interested in everything to do with honey bees.

Hill of Birches

My first beehive

When I had finished my school years at Glasgow Academy I studied at the West of Scotland Agricultural College at Auchincruive for my Diplomas in Agriculture and Dairying, so was able to take night classes in beekeeping taught by Mr Joseph Tinsley in the winter of 1944-45.

Bees were very difficult to obtain during the war and had to be ordered around 9-12 months before delivery – bees got a sugar ration, so nobody gave them up easily! However, I eventually discovered a small beehive company in Exeter, J T Burgess & Sons, and I ordered a colony from them (Italian hybrids) for the price of £7.10/-. The bees arrived in Glasgow Central Station on VE Day (8th May 1945) – a very good omen I have always thought. I still remember the hum, the warmth and the lovely scent from the travelling box. Also, the excitement in transferring them to the newly painted WBC hive. They thrived and made 30lbs of honey that year which was controlled at 2/9 lb extracted.

I began to keep a diary of my newfound hobby!

After leaving Auchincruive and during the first few years of my beekeeping I was working for Joseph Rank Ltd. (father of J. Arthur Rank, the film producer) advising farmers what to feed their livestock and trying to sell our particular "Blue Cross" animal foods. I was provided with a car, only I would have had to pay extra if I wanted it to be fitted with a heater kit!

Photograph: Alex D Henderson, by kind permission of Scotlands' Rural College (SRUC)

Life in digs was now the norm with weekends spent on beekeeping tasks. My job took me around several different areas – the Lothians, Lanarkshire and West Scotland, until one day in 1952, I was asked by my employers if I could take on Perthshire! What luck! This is where I had to go with my bees, back to Blairgowrie and the famous raspberries, back to the glens I had loved as a boy.

I quickly moved my lodgings to Blairgowrie and soon after I found the ideal place to set up a small bee farm – the old stable block at Achalader – just a mile or two west of town. There was no electricity but I put that in fairly quickly.

The bees were brought up from Glasgow and split into several new sites near the town. The extractor was moved into the stables. Soon after this however, Ranks asked me if I would move back to Glasgow Headquarters. I asked for a day or two to think about it. On a Sunday afternoon I took a walk up Strathardle and sat on a ridge looking over the heather moors. I knew I could never go back to the city, and so left their employment. I was left in nice lodgings with 50 hives, no car and very little capital.

My father, who was definitely not in favour of "impossible" full time beekeeping kindly took pity. He gave me the funds to buy a small A30 van, and a small amount for expansion. He also said this should keep me going for a while, and if I did not succeed, he could probably get me a job in Glasgow. He added that I might make more as a tram driver!

Achalader Honey Farm

Moulin Moor

I had burned my boats, but that year 1955, turned out to be one of the best honey years in living memory. The queen bees were so prolific in their egg laying that I managed to split the hives twice in one season – a rare event indeed. I ploughed my profits back into the business and scrimped and saved every penny.

I gradually built the hives up in number and by a careful policy of selective breeding, developed improved strains of bee. Our bees are dark in colour and have a built-in capacity for hard work even in dull weather. The Strathardle bees have many of the similarities of the old-fashioned Scottish bee. Italian bees, which are still great favourites, are inclined to work only if the weather is warm and sunny and in Scotland these conditions can be rare. I had many disheartening experiences, but survived, and I had a lot of very good luck. It took me some years to get used to stings but then I was more or less immune to them.

Raspberry bush

I met Barbara Clark in 1956 at a barn dance at Laighwood, near Butterstone and we married in 1957.

Many of her family originated from around Clunie in Perthshire and her father had worked in banking, starting in Blairgowrie before serving at Gallipoli during WW1. He was then with the Hong Kong Shanghai Bank in Hong Kong, Peking, Bombay, Calcutta, Penang, Dairen, San Francisco and Ipoh in Malaysia before his retiral. Barbara recalls that while in Dairen (in Manchuria) the Chinese 'puppet' Emperor passed along the street in front of the bank where they had an apartment upstairs but those inside were not permitted to look down on him from above and had to keep away from the windows. Her father was interned by the Japanese after Pearl Harbour but the rest of the family managed to find a ship sailing to San Francisco where they lived in Carmel for the rest of the war.

Dairen (today Dalian) branch of HSBC

Photograph by G Warren Swire, image courtesy of John Swire & Sons Ltd and Special Collections, University of Bristol Library (www.hpcbristol.net)

Barbara studied at Edinburgh Art College, where she recalls (vividly) sketching Sean Connery in the life drawing class, and afterwards became a secretary. After our marriage, she kept her secretarial post at Smedleys soft fruit factory in Blairgowrie (raspberries and strawberries being one of the region's biggest exports) for a year or two, helping me in her spare time to extract honey and assemble more hives.

Our Wedding Day

We were very lucky to obtain the gardener's cottage at Achalader, for a very small rent and began our lives there together. Fiona was born in 1960 and Ian the following year, with three more daughters, Catriona, Sarah and Valerie to join us in the years ahead.

Gardener's Cottage, Achalader

With low expenses, I soon repaid my father and we spent everything else on more hives, the numbers grew and we took on our first employee, Bob Barrie. My father had by this time admitted defeat and would often offer tips and advice in order to be of assistance. He had noticed an article in the paper regarding possible insurance against bee rustling which he drew to my attention. On another occasion, when he had been opening a box of Yo-yo biscuits, it occurred to him that a honey Yo-yo might be quite tasty. He happened to know the Yo-yo biscuit maker and wrote to him to say that his son might contact him. I dutifully did so and offered some samples should they consider it a possibility. The company came back to me and said that after some thought it would probably be unlikely that they could use honey, as a biscuit, by nature, should be crispy, not moist. Of course, nowadays, there are biscuits of every conceivable sort (maybe even honey).

In 1962 we bought Glenardle at Bridge of Cally, north of Blairgowrie, a large property with plenty room for a purpose-built bee farm…

CHAPTER 2

My Beekeeping Diary 1945-1956

1945

Hill of Birches, Thorntonhall, Lanarkshire

I was very keen on the old WBC hive and almost bought 20 to be made locally by joiners but at age just 17 and as my future was uncertain I gave up on the idea.

I took my bees to the heather moors at Loch Thom above Greenock. That summer was fairly good and the bees were prevented from swarming by culling the old queen and leaving one cell.

At the end of the season I took my eight honey sections into Glasgow and sold them for £4. I was in business!

Crop: 35 lbs

1946

My bees wintered well and were very busy in spring on all the fruit trees. They were prevented from swarming but gathered only about 20-30 jars of honey. For some reason the clover did not yield well although there was plenty of it.

Wiltshire Camp (myself middle row right)

Towards the end of the year I was called up and entered basic training in Wiltshire. The bees were taken to the new Vale of Leven Industrial Estate site. The Army for me was quite a pleasant change from the heavy farm work at Barrance Farm, Newton Mearns spent hay-making, thinning turnips and harvesting oats and wheat.

It was very wet in May but this is no use for a good clover honey flow later on. A wet April, I have found, seems to give a good clover honey flow. The weather that season turned out to be poor for honey throughout most of Scotland.

1947

Vale of Leven, Dunbartonshire

As I was down in Wiltshire near Salisbury with the Army until July, I saw the bees only once or twice when on leave. For the first time (and without my knowledge) the hive was run for sections. No method of swarm control was used and the bees naturally swarmed. This swarm was hived in another WBC hive. About 30-31 sections were taken off the parent hive in early September and more room was given to the swarm which sadly needed it. They were then left to the tender mercies of the estate gardener!

The weather seems to have been very poor generally until August and little honey was obtained except in heather areas where the crop would be quite good.

The Vale of Leven would have given a good honey yield under consistent management.

1948

I returned home from the Army on May 28th, having served in Palestine with the Military Police [Ed. One morning he put his foot in his boot and encountered a scorpion – he still bore the scar], and on June 8th I brought the 2 WBC hives back to our new home, the Hollies, in Albert Drive, Pollokshields in Glasgow.

I began my three year course at Auchincruive Agricultural College.

I bought 4 Langstroth hives and transferred the 2 WBC colonies into them by cutting combs from the standard frames and tying them into the larger Langstroth frames.

The hives were prevented from swarming by periodic removal of queen cells, but a nucleus was made up with one queen cell just before taking the 3 hives to the heather, near Loch Thom, south of Greenock. This queen mated safely (with a dark drone) and thus 3 hives were established.

Only 21 sections were obtained this year – all from very early blossom in April-May. The weather after spring was poor for the rest of the season.

Honey rose to 8/6 – 10/- per section retail!

The Farm, West of Scotland Agricultural College, Auchincruive.
Photo Alex D. Henderson by kind permission of Scotland's Rural College (SRUC).

**The last general draft 1948
(pre-National Service)
Alamein Camp, Jerusalem.**

**60 Jewish civilians killed in an explosion of 2 army trucks.
The terrorists, disguised as Police, drove off in stolen Police cars.**

My first three Langstroth hives at 'The Hollies', Albert Drive.

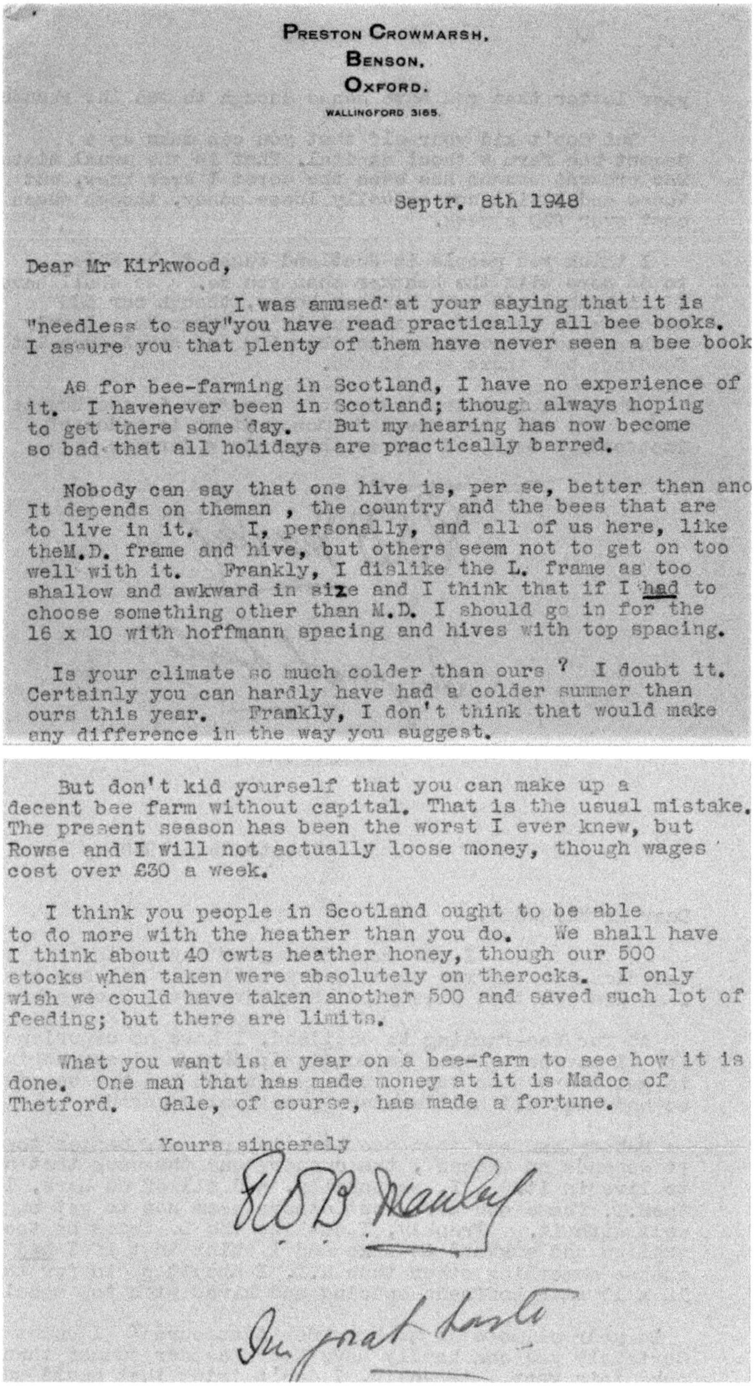

PRESTON CROWMARSH,
BENSON,
OXFORD.
WALLINGFORD 3165.

Septr. 8th 1948

Dear Mr Kirkwood,

I was amused at your saying that it is "needless to say"you have read practically all bee books. I assure you that plenty of them have never seen a bee book

As for bee-farming in Scotland, I have no experience of it. I havenever been in Scotland; though always hoping to get there some day. But my hearing has now become so bad that all holidays are practically barred.

Nobody can say that one hive is, per se, better than ano It depends on theman , the country and the bees that are to live in it. I, personally, and all of us here, like theM.D. frame and hive, but others seem not to get on too well with it. Frankly, I dislike the L. frame as too shallow and awkward in size and I think that if I had to choose something other than M.D. I should go in for the 16 x 10 with hoffmann spacing and hives with top spacing.

Is your climate so much colder than ours ? I doubt it. Certainly you can hardly have had a colder summer than ours this year. Frankly, I don't think that would make any difference in the way you suggest.

But don't kid yourself that you can make up a decent bee farm without capital. That is the usual mistake. The present season has been the worst I ever knew, but Rowse and I will not actually loose money, though wages cost over £30 a week.

I think you people in Scotland ought to be able to do more with the heather than you do. We shall have I think about 40 cwts heather honey, though our 500 stocks when taken were absolutely on therocks. I only wish we could have taken another 500 and saved such lot of feeding; but there are limits.

What you want is a year on a bee-farm to see how it is done. One man that has made money at it is Madoc of Thetford. Gale, of course, has made a fortune.

Yours sincerely

R O B Manley

In great haste

The beginning of my correspondence with R.O.B. Manley.

PLEASE ADDRESS ALL COMMUNICATIONS TO MARLBOROUGH.

A. W. GALE,
BEE-FARMER.

40 APIARIES AT
MARLBOROUGH, SHALBOURNE,
BEDWYN, COLLINGBOURNE, ETC.

TELEGRAPHIC ADDRESS:
"GALE, MARLBOROUGH."
STATION · MARLBOROUGH G.W.R.

Marlborough,

Oct.5 1948

Dear Sir,

Thank you for your letter of the 1st. inst.

I have no wish to interfere in any way with your plans nor to try to persuade you against yourideas; it makes not the least difference to me whether you use Langstroth, M.D. or B.S. frames. But I don't like to see people fail, even though it is none of my business, & from my own experience I do think that you would be very well advised, before spending much money on M.D. equipment, to try a few of them for at least 2/3 years before investing any considerable sum of money in this size.

It is my own experience that bees do better if cramped down in a small space for the winter, & given additional room in spring. I doubt not that theoretically, the M.D. advocates can win every time; I confess that I myself don't understand why it is that bees seem to give more honey on B.S. frames than on M.D., but they do. If M.D. frames were in my view a better size for honey production, I should not stick to B.S.; but as I wrote before, I tried out both sizes side by side, in several apiaries, over a period of years, & the B.S. always won easily. Nor do the crops secured by M.D. enthusiasts strike me as particularly good; in the very last letter Manley wrote me, a week or two ago, he said " Of course, you ~~will~~ always get a lot more honey than we do"---he seems quite resigned to it! And he once wrote me that any man who claimed to average more than 56 lbs. p r stock in the best year was a liar. Well, I must be a hell of a liar at that rate.

Another point----in the next 100 years there is every probability that you will either die or give up your bees, which will then have to be sold. On B.S. frames you can sell them easily; but just try selling them on M.D.s. You just won't get a buyer. Nor, if you want to sell a few stocks at any time, will you find it easy to do so on M.D. frames. There is a lot to be said for doing the same as the Romans do, when in Rome.

PLEASE ADDRESS ALL COMMUNICATIONS TO MARLBOROUGH.

A. W. GALE,
BEE-FARMER.

40 APIARIES AT
MARLBOROUGH, SHALBOURNE,
BEDWYN, COLLINGBOURNE, ETC.

TELEGRAPHIC ADDRESS:
"GALE, MARLBOROUGH."
STATION · MARLBOROUGH G.W.R.

Marlborough,

Oct.12 1948

Dear Mr.Kirkwood,

I don't know it all, by a h...of a long way, but if you are starting up & I can give you any little tips, let me know. This applies just the same if you buy your bees elsewhere! I bought my little knowledge the dear way---doing things wrong---& as Mark Twain used to say "Experience is a good teacher but the tuition is rather high"

Yours sincerely

A W Gale

A helping hand from A.W. Gale, 12th October.

PRESTON CROWMARSH,
BENSON,
OXFORD.
WALLINGFORD 3165.

Novr 8th 1948

Your letter was taken out to the office for Rowse to answer as he does most of that work, and I suppose he threw your letter into the W.P.B, and as luck would have it, this was clearing out day and the whole lot went on the bonfire. And I cannot remember your address, so I am asking your bank to forward this.

Thanks for your cheque 21/6 for the book. It has not yet come to hand, though I am expecting it any day. Fabers said it would be ready before this; but you can't depend on anything these days to be delivered to time, except tax demands.

I will send the book immediately it arrives. I shall have the first consignment that is printed. And will be very pleased to write my name in it for you.

I don't think these blessed books are worth the money myself; but a lot of people seem to. I hope you will.

I believe Rowse answered your letter about hives, did he not. I think you are very wise myself. Prices will fall before long.

Yours sincerely

R.O.B. Manley

This letter appears to refer to *The Practical Bee Guide – a Manual of Modern Beekeeping* (1949) with J.G. Digges.

Note

For further information on Robert Orlando Beater Manley (29 May 1887-–14 September1977) please refer to his books on beekeeping listed in the Bibliography.

For further informaion on A.W. Gale of Marlborough (1900–1969) please refer to the Bibliography.

1949

352 Albert Drive, Glasgow/Loch Thom, Renfrewshire

The hives were examined periodically, and queen cells removed but one examination was missed, and the bees sent out a small swarm a week later (June 18th 1949 – in Pollokshields).

I united the swarm to the main stock when virgin was mated and laying (and clipped).

The bees were taken to the heather on July 28th after removing and extracting 15 Langstroth combs, which gave 48lbs pure lime honey – dark green and most distasteful looking.

The bees were brought back from the heather on September 19th and all the full depth supers were practically fully sealed. 108lbs of heather honey were pressed out giving an average hive production for the year of 53lbs – my highest so far.

This year was the best for a long time and seemingly not a few hives in good areas gave over 200lbs. The weather was almost ideal, especially in England.

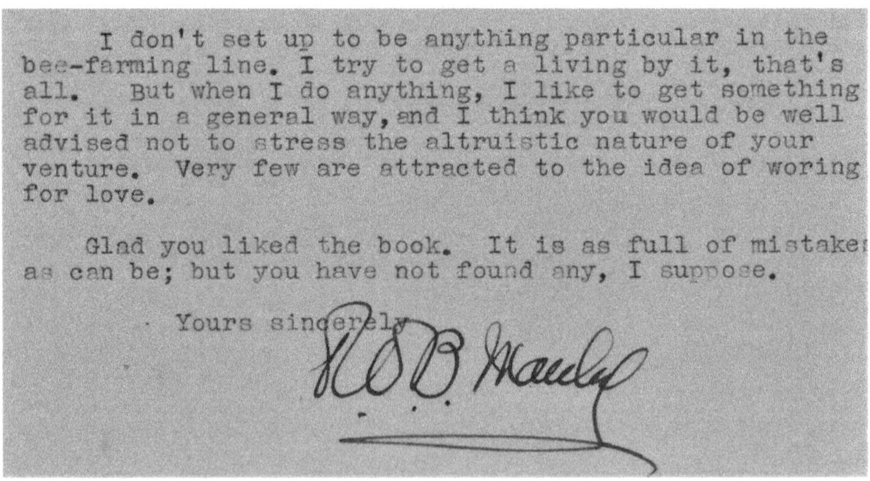

I don't set up to be anything particular in the bee-farming line. I try to get a living by it, that's all. But when I do anything, I like to get something for it in a general way, and I think you would be well advised not to stress the altruistic nature of your venture. Very few are attracted to the idea of woring for love.

Glad you liked the book. It is as full of mistake as can be; but you have not found any, I suppose.

Yours sincerely

R.O.B. 30 January.

1950

West Yonderton, Bridge of Weir

My 3 hives were taken to West Yonderton Farm, Bridge of Weir, Renfrewshire to be looked after, during my stay with A W Gale in Marlborough, Wiltshire.

The day was miserable with rain and cold high winds – I felt some misgivings as I could see no forage and the hives were light.

A.W. Gale (left).

'Brum' and John Madgwick.

A.W. Gale's Honey Farm, Marlborough.

The Old Mill at Werg.

One of A.W. Gale's yards near Yatesbury (grass before cutting).

A.W. Gale's home.

When I got back from Wiltshire, however, in early July, I found that the supers all contained 7-8 combs of honey, and that all 3 colonies were building queen cells galore.

Did the honey come from hawthorn, sycamore or early clover? It could hardly have been from the latter.

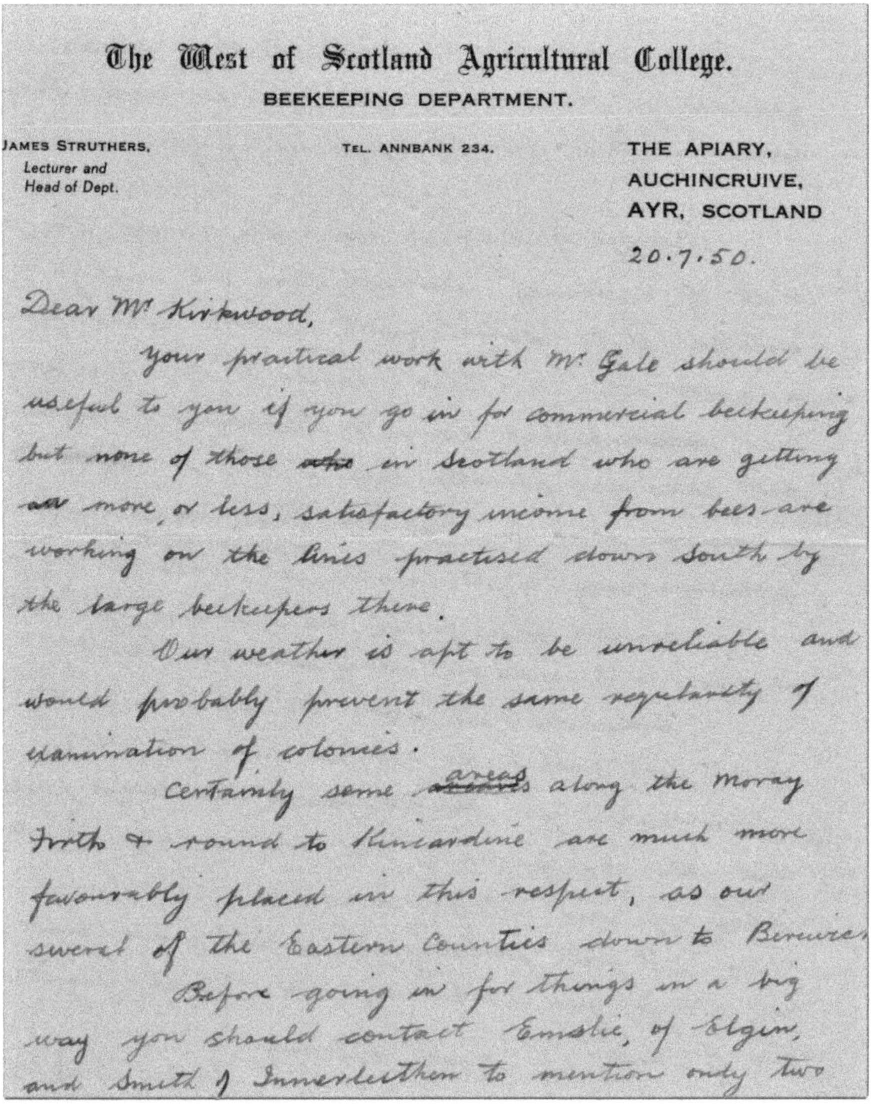

The West of Scotland Agricultural College.

BEEKEEPING DEPARTMENT.

JAMES STRUTHERS, Lecturer and Head of Dept.

TEL. ANNBANK 234.

THE APIARY, AUCHINCRUIVE, AYR, SCOTLAND

20.7.50.

Dear Mr Kirkwood,

Your practical work with Mr Gale should be useful to you if you go in for commercial beekeeping but none of those ~~who~~ in Scotland who are getting ~~an~~ more, or less, satisfactory income from bees are working on the lines practised down South by the large beekeepers there.

Our weather is apt to be unreliable and would probably prevent the same regularity of examination of colonies.

Certainly some areas along the Moray Firth & round to Kincardine are much more favourably placed in this respect, as are several of the Eastern Counties down to Berwick

Before going in for things in a big way you should contact Emslie, of Elgin, and Smith of Innerleithen to mention only two

Continued over leaf.

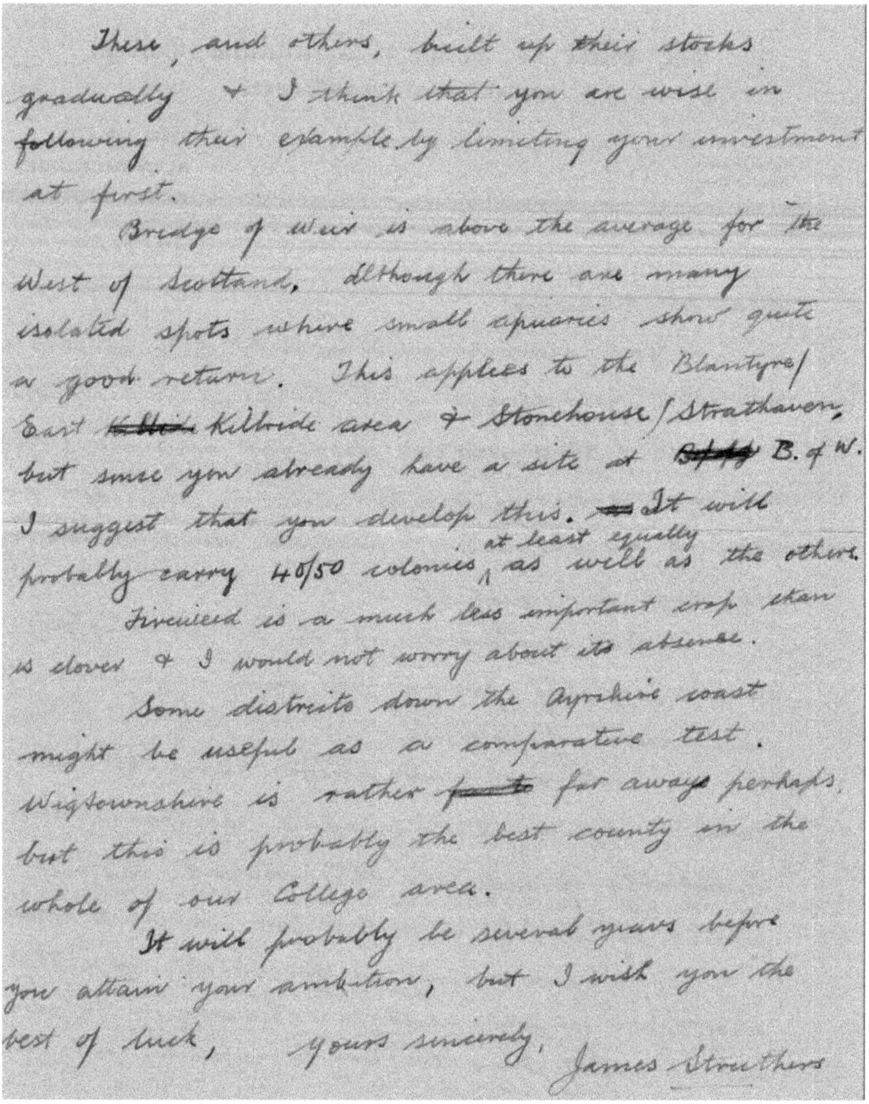

These, and others, built up their stocks gradually & I think that you are wise in following their example by limiting your investment at first.

Bridge of Weir is above the average for the West of Scotland, although there are many isolated spots where small apiaries show quite a good return. This applies to the Blantyre/East Kilbride area & Stonehouse/Strathaven, but since you already have a site at B. of W. I suggest that you develop this. It will probably carry 40/50 colonies at least equally as well as the others.

Fireweed is a much less important crop than is clover & I would not worry about its absence.

Some districts down the Ayrshire coast might be useful as a comparative test. Wigtownshire is rather far away perhaps, but this is probably the best county in the whole of our College area.

It will probably be several years before you attain your ambition, but I wish you the best of luck,

Yours sincerely,
James Struthers

James Struthers, Head of Beekeeping, WSAC, July 1950.
By kind permission of Scotland's Rural College (SRUC).

Heathland Apiaries.

The weather became very poor however after July 10th and it rained almost incessantly right on until September with only a few fair days. The bees did practically nothing at the heather but a breeder queen from Mr Gale and a 1950 queen of my own were successfully introduced.

The weather became very poor however after July 10th and it rained almost incessantly right on until September with only a few fair days. The bees did practically nothing at the heather but a breeder queen from Mr Gale and a 1950 queen of my own were successfully introduced. Nevertheless, 93lbs of flower honey and 6lbs of heather was extracted and hives were increased from 3 to 4. Very little honey was obtained in any part of Scotland, or England (according to Mr Gale).

In July, using the money I had saved while in the Army, I ordered 55 Smith hives from Lees and decided to call myself the "Heathland Apiaries"!

The hives arrived on August 23rd at Albert Drive and I spent a great deal of time after that just putting them together.

6 dozen jars of flower honey were sold at 3/3d jar – the wholesale price common in the north at this time.

"Savernake" herd of pedigree Wessex pigs.

"Savernake" herd of pedigree attested Ayrshire cattle.

A. W. GALE

Church Farm,
Mildenhall,
Marlborough, Wilts.

Phone—Private: Marlborough 159.
Office: Marlborough 385.
Grams—Gale, Mildenhall, Marlborough.
Station—Marlborough (Western Region)

July 28 1950

Dear Athole,

I got quite a shock when I saw your imposing notepaper & realised that we had had such an important visitor this summer!

You went back at just the right time to escape the heavy work, you artful blighter; we are now in the throes of preparing the bees for Exmoor, & it is a hell of a rush. We have sent down two loads this week, & hope to send 4 more next week & 4 the following week. Bees here are almost starving; we have had nothing but wet, cold or high winds & the sup are far lighter even than they were 3 weeks ago. I doubt whether we shall get 4 /5 tons this time, apart from any heather; in fact, 5 tons would very agreeably surprise me. It seems the same everywhere in the South; on Exmoor no-one had a pound of honey, & I have heard from some other commercial folk that they are the same.

Now as regards the bees, I don't know how prices will go next year; they will certainly be no higher than this year & probably less. In any case, I will supply you with 50 stocks or nuclei, at one-third below the retail price whatever that may be. What I suggest is sending you stocks with Buckfast queens; they will give you a season's work but should not be carried later than the following spring. These will be queens reared last July, introduced to his honey stocks by Br. Adam this March/April, & used this year; next spring they will be exchanged for 1950 queens wintered in his mating nuclei & I shall get them in early or mid-May when he starts his queen-rearing. We have quite a lot of queens in our honey stocks now, that I bought from him in May 1949 & they are still going strong, but we are replacing them as we get the young queens available. Of course, if you would rather have RFs or SH queens, or hybrids, you can have them; but I think these Buckfast queens would probably suit you.

With kind regards
In haste

A.W. Gale *in haste*, July 1950.

1951

The 4 Langstroth hives were taken back to the new apiary site at West Yonderton Farm in May after a new hut was erected and fenced in. On 6th May the 50 Smith hives (without supers) were moved down to the site by a Mr Hastie and myself.

15 French packages of bees arrived from Steele & Brodies on May 12th and were eventually all hived and fed. Unfortunately, the queens were not confined to small cages and it was not surprising that 6-7 of the swarms left their hives the next day – while I was away with my friend, Eddie Millington, enjoying myself at Loch Lomond!

[Later – It is thought that 3 swarms left completely and took up residence in empty hives about 1 mile away, on a neighbouring farm!]

On Monday 14th May, when I should have been making cheese at Auchincruive, I went down to the apiary and spent all day there trying to split up one massive swarm, composed of 4-6 packages, which had all united, on a fence post. I had one h… of a time trying to separate them, but was only partly successful and had to hive the major part of this swarm in one two-story hive.

First Commercial Yard, West Yonderton, Bridge of Weir.

Eddie Millington and me.

Needless to say, on my way to Ayr that evening my Matchless 350cc motorcycle broke down hopelessly at Kilwinning, leaving me a very dejected creature by evening.

Investigation revealed on May 17th that 10 packages (now of varying size) were alright and building up, but several were very weak indeed.

On 20th May two of the swarms were found to be dead, having been robbed out completely and killed. I found on 23rd May another one had met the same fate and by 26th May my apiary was sadly deflated, consisting of 1 mammoth stock, 5 decent packages, 1 weak package and 2 good Langstroth stocks (the other 2 being queenless).

I decided on 27th May to use the mammoth stock for drawing combs so put on another super and fed heavily.

By 30th May only 3 packages were found to be left at a decent strength, and I was about giving up beekeeping but the mammoth stock made up for the others in a way, having fully drawn out 48 frames of foundation in 18 days!! But I found a dozen queen cells in this stock. On 1st June I dequeened the big stock and made up a nucleus with the queen.

I went to the bees in the afternoon of 3rd June (having been on duty at Auchincruive in the morning) and grafted 30 eggs or young larvae to a cell building stock on Langstroth frames. These larvae were from a Gale Buckfast Italian strain.

That day I got a sting on the tip of the nose and did it make me cry!!

My Matchless 350.

SDDH Course, Auchincruive 1951 – I'm at the very back row second from right.
Photograph attributed to George Crawford, by kind permission of Scotland's Rural College (SRUC).

Disappointingly, on 6th June I found that only half a dozen of the cells had been accepted probably because the stock was not strong enough to look after more.

On the way back to Ayr that evening I skidded badly on a wet road outside Johnstone. I nearly came to grief but somehow managed to keep the Ariel motorcycle under control.

On June 9th I split the mammoth stock into 9 – still leaving it strong enough for cell building! The French bees were very vicious however and difficult to handle. Then I grafted more cells to both the large stock and the (1) Langstroth cell-building colonies.

By June 16th I decided there was something wrong with my grafting methods because only 4 queen cells were found in the Langstroth hive and none in the large hive – because of queen cells on their own brood. I fed some honey (lime) to the big hive and tried again with more larvae.

I released the nuclei which had been made up about a week earlier on June 17th but found that one was quite empty – how?

On 20th June 3 8-frame stocks arrived from Lees. The bees were nicely coloured and had wonderful tempers – in fact one could handle them easily without smoke or gloves – on a warm day. Would they work? That was the problem.

By kind permission of
Scotland's Rural College (SRUC).

I made up another nucleus on June 22nd and introduced 3 new queens sent by Mr Gale at 15/- each. On June 23rd I found the Langstroth hives to be hopelessly vicious so left them alone (one with a virgin) and grafted larvae to one of the Lee stocks – on 6 frames only.

On 29th June I found that three queens, bred entirely by myself, were laying satisfactorily in the nuclei. This gave me a great feeling of "a job well done" but I found mating queens to be a problem later.

At this date the French packages were covering 8-10 frames and were becoming quite strong.

Another 5 8-frame stocks arrived from Lees on 30th June. I had to saw down the long lugs as before, but the clover was yielding well and the bees were quite amiable altogether. Decided that the Lee Italian strain was quite "my cup of tea" if they would work.

By July 8th I had 46 nuclei/hives, all with queen cells or queens. The clover was yielding extremely well and strong colonies would have obtained a good surplus, but my hives were mostly too weak, though I obtained 8 nice sections from one Lee stock. I found on 16th July that there was a small swarm in an empty Smith hive, over at the farm, so moved it on three frames, to my 50th hive. Left in on the same stand, till moved to heather.

On July 21st many nuclei were still queenless and I began to despair of the poor weather, but I went with Eddie Millington to the heather near Loch Thom and finally found a site for the apiary in a thin wood at Gorshangan, home of a Mr Hill.

I gave a comb of young brood to each nucleus where the presence of a virgin was doubted on my July 22nd visit. There were now 21 queens laying altogether.

I caught another small swarm near the farm and hived it at the spot in my 51st hive. By July 30th 24 queens were laying. The weather was quite good but windy.

First Commercial Heather Apiary of 40 colonies at Garshangon, Loch Thom, Renfrewshire.

I found a pleasant surprise awaiting me on August 9th – another 12 queens noted – but I killed one accidentally when opening the hive. 2 old queens were missing.

By August 5th 34 queens were laying. Lime trees were yielding well to strong stocks.

When I visited on August 11th hives were equalised to some extent and the hives were locked with perforated zinc over the doors for moving the next day. Floors were stapled to brood chambers.

Next morning on 12th August the hives were moved to the heather. It was cold, miserable, none too promising, but I got home at 6.30pm.

The week ending August 19th was cold, damp, with only one good day. No more queens were mated. Did a bit of equalising. Some bees short of stores.

36 queens were laying by September 2nd but I decided that the site was much too exposed, as the wind seemed to blow hard through the wood all the time. As one day this week was very good some of the queens would be mated but not yet laying.

On September 16th I locked the bees in for their return. As it was cold and wet, I finished this job by 6pm. Several more queens were laying.

The next day the hives were taken back to Bridge of Weir. Most were heavier than when they went. A few were considerably heavier. Altogether it was worthwhile to move the bees.

Syrup could not be prepared for feeding until about September 28th as the first permit was lost and there was a considerable mix-up with the order.

Feeding was finished only by October 17th, most stocks having been given 14-20lbs of syrup. 32 stocks were ready for winter, most covering only 8 frames but some few covering 11–12 frames.

The number of combs in the hive was reduced to the number which could be covered by the bees – to reduce any mould growth and to make the hive fit the colony.

Another 40 gallons of syrup was fed to the bees – at a cost of 10/- per gallon for sugar. (This syrup was extra thick, 1 cwt sugar: 6 gallons water – made in the bath at home)!

The total honey crop for the season was 20lbs from one stock and 10 sections from another. All others were split up and had no hope of giving a surplus.

The stock which gave 20lbs was on 8-frame stock of Italians from Lees of Uxbridge. It had removed from it 4-6 combs in addition – at different times to assist storing nuclei. As this stock was therefore only about one third of full strength at the clover flow, a reasonable average expectation could have been about 45+lbs/full stock.

Altogether in the year 56 visits were made to the Bridge of Weir apiary – some from Ayr and some from Glasgow. An estimated milage for the motorcycle(s) was 2,100 – in bee work alone.

1952

The winter was mild and good for bees, but I lost 4 stocks – one from starvation and three from queenlessness or drone-laying queens. The stock which stored was one of the strongest. It had eaten all the honey at one side of the hive and had been unable to reach the other side (where there was honey left) because of a cold spell. 2 other colonies became queenless later. There were therefore 26 colonies on 13th April, and I resolved to split some of them later in a (vain) attempt to reach 50.

On 21st April I started feeding with thin syrup (28lb: 8Gal) and by the 26th there were 2-4 frames of brood in most hives.

By 5th May one or two stocks were almost ready for supers but it was not until the 15th that they were actually required – some stocks by this time being quite crowded for space.

Half the colonies were supered and working well on 3 frames in the supers by 18th May. It was very hot indeed and this certainly helped the rate of egg laying. Drones were now beginning to emerge from their cells, but queen cells were still absent.

On 1st June preparations were made for artificial queen rearing but the weather turned cold. Indoors I just finished fitting 615 sheets of foundation into wired frames.

A few artificial queen cells were ready by 11th June so 8 artificial swarms were made up – from colonies which were now fully occupying 20 frames. On 12th June one French colony was found to have brood in 15-16 frames.

By 15th June 43 hives had been set up, and the best stock (unsplit) was given a third (full depth) box of frames. One artificial swarm at this date had expanded so rapidly that it required a super. Some of the parent stocks however contained chilled brood and eggs – a result of the loss of flying bees.

On 26th June there still wasn't a drop of surplus honey – and very little in the brood chambers, but on 29th June the first young Lee strain Italian queen was laying, clover was yielding well and things looked better all round.

This fortnight (1st July and last week of June) Glasgow Fair Week, was really marvellously hot and the bees did very well. Several more queens began to lay and some colonies drew to frames of foundation inside a week. The bees were very busy towards the end of this fortnight in spite of a strong (warm) east wind.

The weather turned poor after this however and much honey was consumed. But by 13th July 39 queens were laying and I had successfully introduced 3 pure Lee Italian queens (to young bees). The first queen cells appeared on 12th July quite possibly due in part to the wet weather in the last week or two.

On July 20th queens were first drawn into one box and an attempt was made to sort out former combs into the supers (CD). This was very difficult however due to the fairly large number of Yorkshire-spaced frames mixed up with the Hoffmans.

The temperature rose to 71° in the week ending 26th July and I estimated that there were 80 frames of unsealed honey there to be taken off later. Several queens were missing however – in stocks which had attempted to swarm. Probably because they were clipped and got lost in the grass.

All virgins that were ready were mated in the glorious weather – aided perhaps by wooden boards which acted as markers outside the different hives. A little of the clover honey was now sealed but 20 stocks were still not supered at all.

I removed 40 frames of clover honey – some unsealed, on August 2nd. Then I proceeded to lock the bees in for the moving and as it rained steadily I was really soaked.

On August 3rd 45 colonies were taken to the heather by Mr Mackie, Bridge of Weir, myself, another man, and a boy.

This year we took them to a new and excellent site – Diffany Farm, Kilmacolm. The apiary was arranged 50 yards from the moor at the end of the farm road – a road which went nowhere in particular – hence quietness was assured.

Excerpt from article by AMC Kirkwood in the *British Bee Journal* Honey Market Report dated 12th January 1952

...The most popular form of honey with the consumer is undoubtedly the 1lb heather section, but at 5s. to 6s. sales are restricted, and so more of the crop today is sold in 1lb jars at 4s.6d. or so. Clover honey is becoming increasingly difficult to sell due to the public preference for heather, and it is doubtful if sales will rise until the price is down to 2s. 9d.

Little honey is sold in ½lb jars but this is a mistake as the smaller pack could be retailed quite economically at 2s. to 2s. 6d. – a price well within the purchasing power of most housewives.

As far as can be ascertained most retailers make 20-25% on sales of honey, but many make a lot more, especially in good seasons when the smaller beekeeper does not like to store his crop and sells it instead at less than the average cost of production.

The Scottish Beekeepers' Association has done much good work with respect to this problem but can only recommend minimum prices unfortunately, instead of being able to enforce a minimum "rock-bottom" price.

...I would like to take this opportunity of wishing all readers a very prosperous New Year, full of pleasant surprises and general happiness.

The first fortnight or so at the heather was cold and wet, and bees seldom got out. The rest of August was quite reasonable, but it was never very warm. The bees did exceedingly well however, and the best stocks stored about 40-50lb of fine heather honey. Those 15 stocks which occupied only one box stored no surplus, but the brood chambers were well stocked for the winter. The stock which had done best at the clover was given 3 section supers of foundation. One of these was filled completely but the colony was really past its best and other stocks approaching full strength did much better.

While the colonies were at the heather I went on holiday, but not before extracting 1cwt of clover honey. In doing this I used the steam plane and managed to wreck my copper boiler through letting it cool without admitting air!

I inspected the hives on August 30th and found one quite upside down with the floor half off and the super loose. The bees appeared quite comfortable but were rather vicious. The hive must have been upside down for some days as the grass below was pale in colour. Cattle had got in probably.

The hives were brought back to Bridge of Weir on September 7th as the season had been an early one, and the heather was past.

On the 9th, Porter escapes were put on, and by the 14th all the honey had been taken off – 16 boxes in all – but many combs only half filled.

Excerpt from article by AMC Kirkwood in the *British Bee Journal* Honey Market Report section dated 11th December 1952

...The first National Honey Show to take place in Scotland was held last month in Glasgow. It was a great success and attracted many shopkeepers as well as the general public, and beekeepers from all over the country.

An exhibition such as this will help honey sales immensely, especially if it is well advertised beforehand in the daily papers, and in the journals associated with the fruit, dairy and grocery trades.

Once shopkeepers have seen the exhibits produced at the shows they will be inclined to take more of an interest in honey, and many will ponder over the idea of erecting a small stand similar to those they have seen.

But retailers are busy men and if the beekeeper is handy with tools he should take advantage of the situation, making a simple but suitable stand for the shopkeeper. This would not only increase his own sales, but that of the whole beekeeping fraternity. Only one or two electric light bults and some old tea-chests are required. With fretsaw and paint brush there is plenty of scope for originality.
[*Ed. Why have I not seen this ingenuity in Tescos?*]

One of the new features at the Show was the large amount of cut comb or chunk heather honey. It was prepared for sale in many different forms, but somehow it did not have the same neat look as the sections. It is certainly much easier to produce, but unless a better and stronger package is forthcoming it will never replace the section in popularity.

After uniting queenless colonies there were 40 left to winter. Feeding was begun on 21st September and finished on October 4th. The honey was all pressed out into 28lb tins by October 6th and all bottled by the 12th.

Altogether 20 boxes were taken off the hives and one crate of sections (18 sections and two shallow frames).

The total weight of honey produced in the year was 560lbs – mostly heather and the average price realised was about 3/6 – 3/9 lb.

The deep brood chambers (each with 12 frames) only gave an average of 27lbs each – rather a smaller quantity than the usual 55lb estimated!

During 1952 50 visits were made to the apiaries, mostly on the motorcycle. Approx 160 hours were spent at the apiaries in actual work.

The season was quite good, generally speaking, and most beekeepers were fairly happy with their results, but it was not really warm enough in July and August.

In England, however, a state of drought existed, and yields were very low indeed.

My best colony – the only one which was not severely depopulated, gave about 65lb clover and 35lb heather honey – 100lbs in all.

Bell Heather.

1953

Bridge of Weir

1953 had a remarkable Spring. Right from New Year till March the weather was mild and open with only one or two very short spells of frost. Egg-laying began in January, and stores were used up quickly by March, when feeding was begun (on the 7th). Two colonies were lost by March 7th, one queenless and the other robbed out – a very week one. (Needless to say, combs were not examined at this date where symptoms of queenlessness were very evident).

This left 38 colonies in the middle of March and after this, stores were checked at intervals of 2-4 weeks by lifting colonies and raising inner covers.

No less than 10 colonies were found to be queenless or drone-layers by the end of April and this obviously reduced the number of colonies – only 28 on April 25th.

Sycamore trees yielded heavily from May 1st to 10th and cells were commenced in one or two colonies. Most colonies were occupying two boxes by the middle of May and some required more space. I made up 7 colonies with artificial swarms on 1st June, giving each 6 combs of brood. I also made up a Yorkshire spaced colony for Mr Kerr, Croft-An-Righ, Dolphinton – to whom I sold some equipment.

On Coronation Day (2nd June) I made up another 8 artificial swarms, making 42 hives in all.

On 12th June 15 Lee Italian queens were introduced to parent queenless stocks. The clover began to yield this week. A number of the artificial swarms were building queen cells already. These cells were simply destroyed.

On 20th June I arrived at the apiary and found that the gate had been left open. One hive was badly smashed by the cattle which got in, but the bees inside were in good order. The queen introduced to this hive at the previous visit was laying well! Changed all floors over, so that there was only a ⅜ space below the frames.

Fly fishing catch (brown trout) with younger brother, Ronald on right, father in centre and brother-in-law George Murray on left.

Most colonies still building queen cells – Got very sick of the double brood chambers! Especially as it poured all day.

During the Highland Show week (22-26th June) the weather was very hot (80°) but very little honey was stored.

I started on 26th June to CD the combs and managed in time to get all the Yorkshire-spaced frames into the supers. A good clover flow came on the next day, after some rain.

By 5th July I noticed that the French and Buckfast queens were heading the best colonies. They built no cells, gave bigger colonies and gave more honey than the Lee Italians – but they were also more vicious! The French colonies also seemed to build up later, giving grand heather stocks. There was practically no clover flow at all in 1953, and many colonies went to their heather site (Garshangon) with very few stores. I could not get permission to put the bees to Diffany.

The bees were taken to the heather at 7.30am on Sunday 26th July and I did not see them again till the 16th August when I returned from holiday at Lochaline, Morven. The weather had been glorious however for about a week, and some colonies had done very well – though the Lee Italians did not give any honey in the supers. Some French and Buckfast colonies had 3 deeps fairly solid with honey right to the outside frames in the top box – about 70-80 lbs in all. So had one or two Klee queens bred by myself.

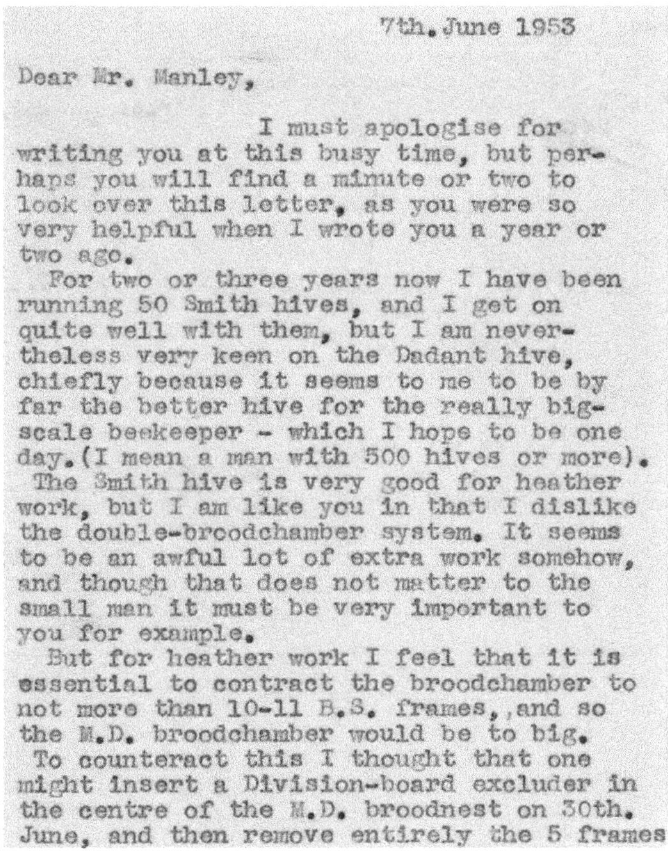

7th. June 1953

Dear Mr. Manley,

I must apologise for writing you at this busy time, but perhaps you will find a minute or two to look over this letter, as you were so very helpful when I wrote you a year or two ago.

For two or three years now I have been running 50 Smith hives, and I get on quite well with them, but I am nevertheless very keen on the Dadant hive, chiefly because it seems to me to be by far the better hive for the really big-scale beekeeper - which I hope to be one day. (I mean a man with 500 hives or more).

The Smith hive is very good for heather work, but I am like you in that I dislike the double-broodchamber system. It seems to be an awful lot of extra work somehow, and though that does not matter to the small man it must be very important to you for example.

But for heather work I feel that it is essential to contract the broodchamber to not more than 10-11 B.S. frames,,and so the M.D. broodchamber would be to big.

To counteract this I thought that one might insert a Division-board excluder in the centre of the M.D. broodnest on 30th. June, and then remove entirely the 5 frames

My letter to R.O.B. Manley (3 pages), 7 June 1953.

2/.

free of brood (and possibly full of clover honey), just before the hive went to the heather. An empty box could then be inserted in the broodnest to take up the space, and this would force the bees up into the super.

Alternatively, if the 5 broodless frames were fairly full of clover honey, they could be left in place when the hive was taken. This also would force the heather honey to be stored inthe super.

I know in the South of England you probably don't place so much importance on the heather crop but up here it is our mainstay.

I have spoken to Mr. Smith several times about the M.D.hive, but he has never used it - so he can't really condem it!

As a matter of fact I have not yet met anyone up here who likes the M.D. hive at all. They all say that the broodchamber is too big (but they don't consider 22 B.S. frames too big in the Smith!)

Apart from this question of heather work I can see nothing against the M.D., apart from selling bees on frames, but even an experiment on a small scale is expensive if you are going to scrap the hives later (I have been trying to sell off a few Langstroth hives for 5 years without success!

Another thing that I have been wondering about for some time is the method of swarm

3/.

control that you give on page 238 of XXxxxxxXxxxx "Beekeeping in Britain".
You say in the book that more experience would have to be gained before you could really recommend it. What do you think of it now? Itwould seem to me to fit in very well with the production of pressed heather honey - simply by cutting out all the combs in the shallows after the colony was brought home.

I know you will probably turn round now and tell me to stick to the Smith hive, but they would always sell easily or would do for producing bees for sale, if I changed to W.D. (There is nothing more popular than the Smith up here just now! - though I heard from my friend McCaig that you had tried them, and given them up again!)

Don't think there is anything else to worry you with just now, and please don't bother to reply at once if you are very busy.

Hope it is a good honey year. If it is'nt I will just have to go on selling cattle concentrates for a few years yet!

Yours sincerely,

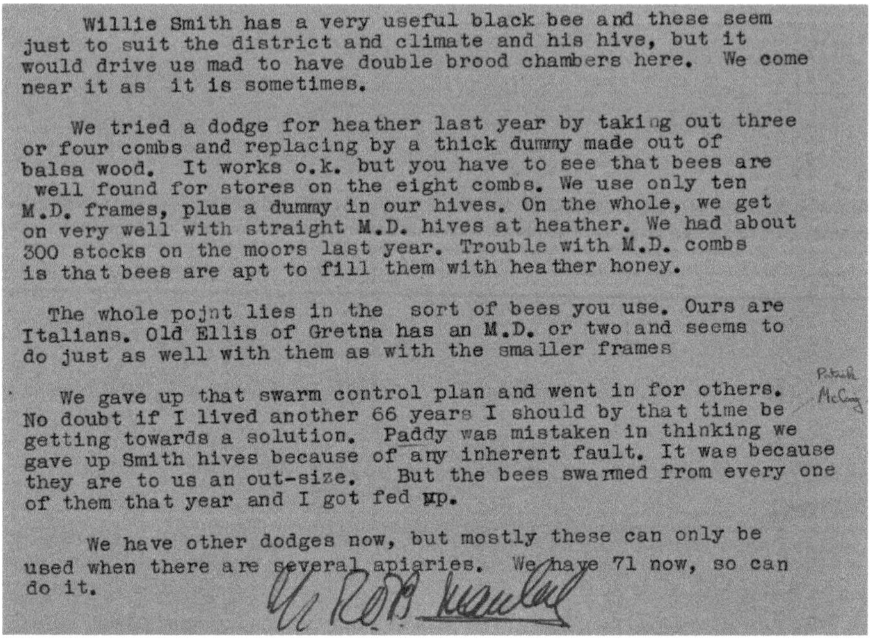

Willie Smith has a very useful black bee and these seem just to suit the district and climate and his hive, but it would drive us mad to have double brood chambers here. We come near it as it is sometimes.

We tried a dodge for heather last year by taking out three or four combs and replacing by a thick dummy made out of balsa wood. It works o.k. but you have to see that bees are well found for stores on the eight combs. We use only ten M.D. frames, plus a dummy in our hives. On the whole, we get on very well with straight M.D. hives at heather. We had about 300 stocks on the moors last year. Trouble with M.D. combs is that bees are apt to fill them with heather honey.

The whole point lies in the sort of bees you use. Ours are Italians. Old Ellis of Gretna has an M.D. or two and seems to do just as well with them as with the smaller frames

We gave up that swarm control plan and went in for others. No doubt if I lived another 66 years I should by that time be getting towards a solution. Paddy was mistaken in thinking we gave up Smith hives because of any inherent fault. It was because they are to us an out-size. But the bees swarmed from every one of them that year and I got fed up.

We have other dodges now, but mostly these can only be used when there are several apiaries. We have 71 now, so can do it.

R.O.B. Manley, 10 June 1953.

The bees gathered very little after the middle of August but were not brought back to Bridge of Weir until 12th September. 3-4 hives fell off the lorry at Bridge of Weir and were smashed to pieces – but the bees were salvaged, and all the hives ultimately repaired.

By the end of October all colonies were reduced to one brood chamber, and most had bees which had been fed 2 Gals of syrup. All honey was in cans by the end of October, but one deep super was left for stores in Spring.

4cwts. of honey was sold in bulk to Moray Firth Foods Ltd @ 4/- per lb. The rest was bottled in ½ lb jars and sold to the shops.

The year was regarded as a very bad one, but I managed 30lbs on Spring count, all heather.

60 visits were paid to the apiary.

1954

Bridge of Weir/Achalader, Blairgowrie

1954 was a very poor season for bees all over Scotland and England. Only in the raspberry areas was honey produced in any quantity and then only from colonies with the queen confined to one brood chamber.

I moved my Bridge of Weir apiary up to Blairgowrie (Achalader Stables) in April after buying in about 30-40 ailing hives from the

Loch Benacally – 76 colonies.

Blairgowrie district. These weak colonies were later transferred to new Smith hives, put together in a great hurry during May and June.

There were therefore about 75-80 colonies at the commencement of the raspberry flow (about June 8th) and these were increased to 100 by artificial swarming.

About 4-5 colonies were left on a single brood chamber for experimental purposes, the remainder being given two.

The difference in the yield of raspberry honey was astonishing. These few colonies each filled about ¾ of a deep super (30lbs) whereas all the other colonies filled the top box with brood – using up most of the raspberry crop in the process!

If the season had continued satisfactorily there is little doubt that the double storey stocks would have put up a much larger crop of clover

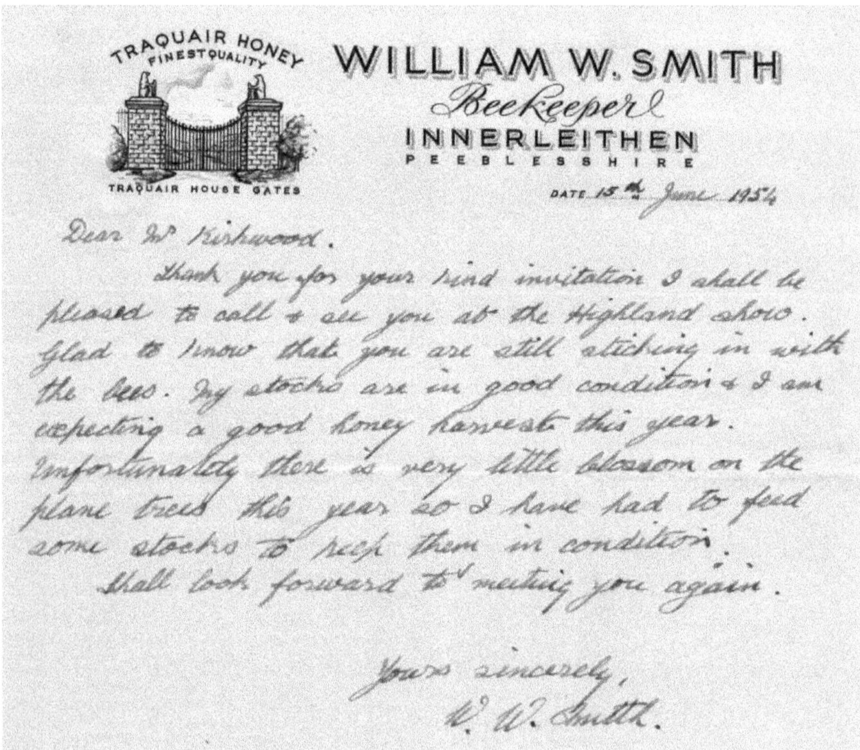

TRAQUAIR HONEY FINEST QUALITY
TRAQUAIR HOUSE GATES

WILLIAM W. SMITH
Beekeeper
INNERLEITHEN
PEEBLESSHIRE

DATE 15th June 1954

Dear Mr Kirkwood.

Thank you for your kind invitation I shall be pleased to call & see you at the Highland show. Glad to know that you are still sticking in with the bees. My stocks are in good condition & I am expecting a good honey harvest this year. Unfortunately there is very little blossom on the plane trees this year so I have had to feed some stocks to keep them in condition.

Shall look forward to meeting you again.

Yours sincerely,
W. W. Smith.

Willie Smith, 31 June 1954.
For further information on William W. Smith (1890–1969) please refer to the Bibliography and Willie Robson's memories of this influential beekeeper in Appendix 2

My 8 oz Heather honey label

Possible designs for labels.

and heather honey but the weather got bad, and it never improved again at all after the end of June. The result was that double brood chamber colonies went to the heather with very little stores and never gave any real surplus, whereas the single brood chamber colonies gave at least 25-30-40 lbs of raspberry honey. Most of the colonies gathered some heather honey for their winter stores and one gallon of syrup was sufficient for autumn feeding.

103 colonies and nuclei were the Autumn count.

The surplus consisted of 221lbs heather honey and 249lbs of raspberry honey.

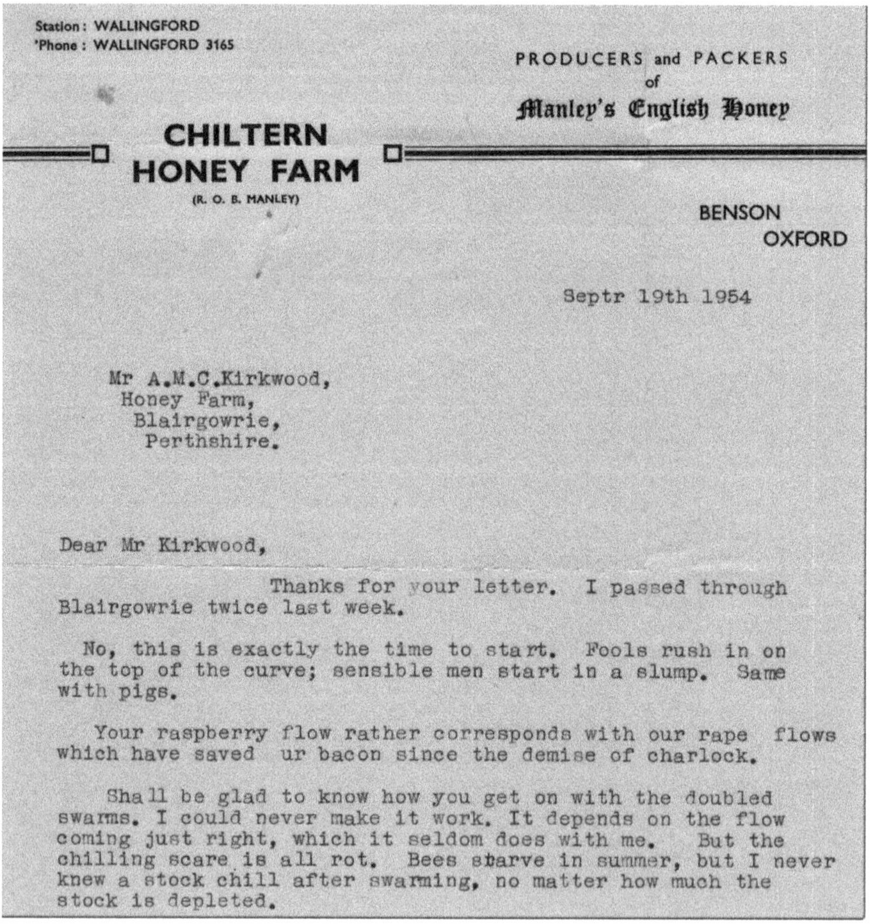

Station: WALLINGFORD
'Phone: WALLINGFORD 3165

PRODUCERS and PACKERS of Manley's English Honey

CHILTERN HONEY FARM
(R. O. B. MANLEY)

BENSON
OXFORD

Septr 19th 1954

Mr A.M.C.Kirkwood,
Honey Farm,
Blairgowrie,
Perthshire.

Dear Mr Kirkwood,

Thanks for your letter. I passed through Blairgowrie twice last week.

No, this is exactly the time to start. Fools rush in on the top of the curve; sensible men start in a slump. Same with pigs.

Your raspberry flow rather corresponds with our rape flows which have saved ur bacon since the demise of charlock.

Shall be glad to know how you get on with the doubled swarms. I could never make it work. It depends on the flow coming just right, which it seldom does with me. But the chilling scare is all rot. Bees starve in summer, but I never knew a stock chill after swarming, no matter how much the stock is depleted.

R.O.B. Manley ,19th September 1954.

1955

Achalader, Blairgowrie

1955 started with a very severe and late Spring. During January to March frost was very severe and there was also a lot of snow.

The result of this weather, which gave the bees no chance for cleansing flights combined with late autumn feeding (which meant a lot of unsealed syrup), was that the bees went down one by one with Nosema in February and March. This was probably due also to the fact that the bees were all in one apiary (Achalader) and got little or no sunshine for a long period.

[There is no doubt that the hives must stand in the sun in such a winter].

The result of all of this was that by April the numbers had dwindled from 100 to 4 effective colonies – surely the most destructive outbreak of disease possible!

I saw at once that I must gamble every penny (literally!) on buying more bees, or there was no point in my continuing in the business.

I therefore started advertising for bees in the local papers, and by the end of May had collected over 60 very good colonies all of which had wintered well on honey. Those with a shallow brood chamber on top were strongest. (Other beekeepers who had fed syrup in Autumn also lost a large percentage of bees). I also ordered 25 French packages from Steele & Brodie, and these arrived towards the end of May.

By mid June I was beginning to feel more like a beekeeper but frost killed the rasp blossom almost entirely, and the flow was very slight indeed.

Then came the most wonderful summer for very many years. Day after day of blazing sunshine.

It was terrible working in the sun in temperatures of 80° or over for some weeks and everything got very very dry but the clover yielded well and during July supers were being filled quite easily in a week.

In July I also went to Stranraer and picked up Mr McCaig's Bee Farm consisting now of about 45 colonies and many spare National and Smith hives. This was all done in a day or two and the bees, when brought to Blairgowrie, were taken straight to the moors at Seafor, high on the hills looking over to Dunkeld and Perth.

Frankly, I am sorry to learn that you have given up your job with Rank to start full-time beekeeping, because I do not think it is going to be possible in future to make a reasonable living from bees in this country. Only yesterday Graham Burtt spent the afternoon & evening with me, together with Dr. Adam of Buckfast who has been staying with me for a few days; we were discussing this same question & we were all three of the opinion that conditions have changed so much during the past few years that it is no longer possible to make a reasonable living from bees alone. I am not referring to the series of bad summers, though this has accentuated it; but I refer to the continual rise in costs of production, the virtual impossibility of getting substantially more for our honey, & the fact that over the greater part of the country the flora is continually getting less for bees, as spraying gets more & more common.

I realise also that all I am writing will fall on deaf ears; you are young, you have your life before you & you have made up your mind to make a living from bees, & good luck to you! But I am sending this warning because I think that it was probably my example that made you feel you could earn a living from bees, & therefore I have some responsibility in the matter which I am now discharging. Only yesterday I was discussing this same matter-----had I started now, instead of 35 years ago, I could never have made a success of the job, I realise that it was just good luck on my part that I succeeded where so many are now failing.

My earnest advice to you---& I know you won't take it---is to get your old job back, or get another, & treat your 200 stocks as a sideline.. You will make a nice little bit in a good year, & in bad years when you make nothing, you will still have your salary, so the wife & bairns won't starve----this is probably looking a head a bit, but doubtless they will follow!

Sincerely yours

A.W. Gale, 31 January 1955.

I agree that the raspberry flow will be a big help, but you wrote that this is too early for the bees to gather much from it, though useful for building up. Nor do I envy you, trying to handle 300 stocks single-handed; but you are young & long hours & hard work won't hurt you. Good luck to you in any case; I like to see a chap try hard & I hope you win through, but frankly I cannot see that you will ever get a reasonable recompense for the work & the skill & forethought you put into it. You won't starve, but neither will you make any worth-while money, unless I am quite wrong. Still, if you like the job, that is worth more than making a higher income at a job you dislike.

I myself am always seeking ways to economise in labour; the trouble is, that so often when one cuts out some work, it later means that still more has to be done, & that is false economy. Of course, when one does the work oneself, it is done better & quicker than by hired labour.

A.W. Gale, 4 February 1955.

The weather continued perfect and all apiaries except Achalader were moved to the heather. Two at Seafor, one at Woodhill, Ballintuim and one at Bridgend, Straloch. The Achalader bees put up a good crop of heather honey without being moved!

The weather had been a bit too dry for the ling to yield well but some colonies got some very nice honey from it just the same. But the Bell heather and late clover were marvellous and supers had to be added and added again.

Some of the best colonies had 4-5 supers filled, well over 200 lbs but Paddy McCaig's bees were quite weak, as he had warned me, through over-swarming and this reduced the average surplus.

But there is no doubt that his drawn combs when given to my bees, helped my production very considerably.

Photographs were taken, and these would easily be mistaken for American, as hives here are so rarely as tall. To sit in the shade of a pine tree and see the bees drift in, in real clouds, was a wonderful experience.

Early in September all the supers were removed from the hives and brought home in the van.

Later a lorry went to the heather and the hives were brought back, two apiaries at a time.

Mr McCaig's hives.

During July and August a considerable quantity of honey was sold, the state of my finances making this a desperate business. But I did not drop the price and obtained 3/- lb for clover in jabs. Heather was sold at 3/9 – 4/- and cut comb heather at 4/6d.

Bell Heather, Straloch.

A large number of National hives was also sold, together with jars and other equipment, all of which helped to put the Honey Farm on a slightly more secure footing!

In addition to my own crop of 4⊠ tons, about 1½-2 tons of other honey (including heather sections) was bought and sold or stored for future use.

The Autumn count was 216 colonies, together with an apiary of 20-40 colonies to be lifted at Stranraer.

Straloch – Bell Heather.

TRAQUAIR HONEY
FINEST QUALITY
TRAQUAIR HOUSE GATES

WILLIAM W. SMITH
Beekeeper
INNERLEITHEN
PEEBLESSHIRE

DATE 9th Aug 1955.

Dear Mr Kirkwood,

Sorry to have been so long in replying to your letter of the 4th June, We were undecided about going to the Highland Show, however we went on the Wednesday.
It has been a great help having Janny with me this year as I have had such a hectic season, My flower honey harvest is exceptionally good, we are having the first rain in a long time, today, it will do heather a lot of good if it doesn't continue too long.
Hope you also have had a good crop of honey to make up for your disappointments.
My wife & I enjoyed our visit to Manley he spent a lot of time in entertaining us, yes they work at high speed but I still would not have the Dadant hive in my apiaries especially in heather honey production.

Wishing you all the best.

Yours sincerely,
W. W. Smith.

1956

1956 was not a good year anywhere. The weather appeared to be quite promising in Spring and there was a slight flow from sycamore, but after May the weather got poorer all the time.

During the raspberry honey season the weather was cold and very wet with frequent thunder showers. The bees worked only on odd days and had only about one good week when the stronger colonies gathered about 25lbs. Many gathered little or nothing on the raspberry flow and a peculiarity of the year was the disinclination of bees to build comb if they could avoid it. This was remarked on by quite a few beekeepers in the area.

The raspberry crop of about 1 ton was extracted by hand in the MG Parallel Radial extractor obtained from Mr Home, Stranraer who gave up beekeeping in April and whose stock I purchased entire. (32 hives of bees and much other equipment).

The bees were brought safely to Hillocks of Gourdie Farm at Essendy and they were there gradually transferred from the very bulky hives into Smiths. The old hives were left, and eventually the berry pickers used most of them for firewood in the Glasgow Fair week, just before I moved the bees to the heather.

The French bees proved again to be the best for the conditions in Perthshire but perhaps the best strain of all was the single colony from my brother, Ian's mother-in-law, Mrs Kerr in Glasgow. The hive was so good that I bred many young queens from it, and hope that they can be perpetuated!

The 6 apiaries were all moved up to the heather by the 20th – 24th July, but unfortunately only the bell yielded (from about 5th to 20th approx.) and the result was that only the Glenshee and Strathardle apiaries (3) did quite well. The other 3 at Benacally, Speid's Wood (Seafor) and Millhole of Achalader did little though they did get something for winter from the ling. On the South side of the hills there just didn't appear to be any bell heather, while at Persie, Woodhill and Straloch there was a very good flow.

Mr Capper and the Austin A30 van.

The 150 colonies in this area gathered 1 ton in about a week of good warm weather. This would have been 2 tons almost certainly (from only 150) if they had been up just a week or two earlier. It appears that all colonies should be shifted about 2nd to 3rd July, but it is difficult to move them all at once in a week or so.

I was fortunate in obtaining 4 Smith hives from Ballintuim House, where they had been left by Mr Telfer who had moved to Persie. In return I shifted his other three or four hives over to Persie for him, after cleaning up the awful mess of honey. The honey had not been taken off in the very good year preceding and the hives were full to the roof and outside the doors, etc and between the walls – but I got a good haul when the bees cleaned up all the mess from the old section crates, etc!!

Throughout the year I was helped by my good old friend, Mr Capper, but unfortunately his wife died in December and he left for Chester to stay with his sister.

The total crop of honey for the year was about 2¼ tons only, but I had about ½ ton left from the preceding season and as the price was good (3/6 clover, 4/6 heather) I had no difficulty in getting £1200 for honey, plus about £70 for honey bought in.

Sales of honey jars were practically nil compared with the preceding good year.

Losses from Nosema were not too serious in 1956, but all combs not in use were disinfected with Acetic acid, and this will be done as a routine measure every spring in future.

At the end of the year more alterations were done to the premises at Achalader (by Bruce the builders) but I spent a lot of time working in the house (which I had moved into in May 1956). The work on the house was important, as my fiancé and I planned to marry at the end of April!

I also experimented for the first time in feeding glucose fondant to the bees as candy and found that 7lb tins on top of the cluster were quite good, provided that the paper was left in the tin with only a 1½" square cut in it. If the candy was removed from the tin it tended to melt and run down into the cluster, killing the bees perhaps.

About a quarter of the bees needed feeding by the New Year and these were all given the 7lb tins of fondant.

The Spring count was approx. 190 colonies + 32 colonies bought from Mr. Home. This made 222, and in the Autumn the count was 273, an increase of 50.

Production Records kept after this date instead of a diary.

Wedding departure!

CHAPTER 3

Bee Farming 1957-1981

1957

Achalader, Blairgowrie

This year I married Barbara and seem to have had no time to keep a detailed Production Record!

Production figures refer to honey actually produced. They include any seed added, but do not include any honey purchased elsewhere.

Bee boles at Pluscarden Abbey, Morayshire – Honeymoon.

1958

Achalader, Blairgowrie

Many thanks for your letter, & I am delighted to know that you are going ahead so well. How much labour are you employing? You must certainly need some help in running 550 colonies, & I have learned by experience that although when you are young & eager you can work almost any hours, you have to pay for it in later years! I crocked up 18 months ago with a groggy heart & have done practically nothing with the bees since then, since I cannot lift anything; there still seems to be heaps to do, office work, planning new farm buildings etc., but I take off quite a bit of time for fishing.

Letter from A.W. Gale, 5th September.

Straloch.

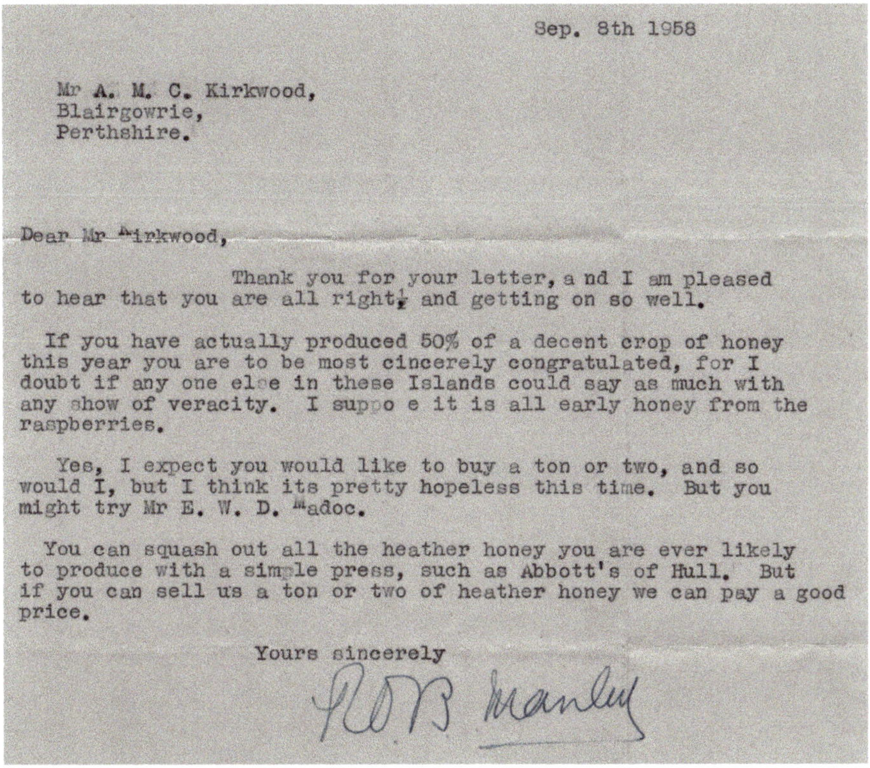

Sep. 8th 1958

Mr A. M. C. Kirkwood,
Blairgowrie,
Perthshire.

Dear Mr Kirkwood,

Thank you for your letter, a nd I am pleased to hear that you are all right, and getting on so well.

If you have actually produced 50% of a decent crop of honey this year you are to be most cincerely congratulated, for I doubt if any one else in these Islands could say as much with any show of veracity. I suppose it is all early honey from the raspberries.

Yes, I expect you would like to buy a ton or two, and so would I, but I think its pretty hopeless this time. But you might try Mr E. W. D. Madoc.

You can squash out all the heather honey you are ever likely to produce with a simple press, such as Abbott's of Hull. But if you can sell us a ton or two of heather honey we can pay a good price.

Yours sincerely

ROB Manley

ROB Manley, 8th September.

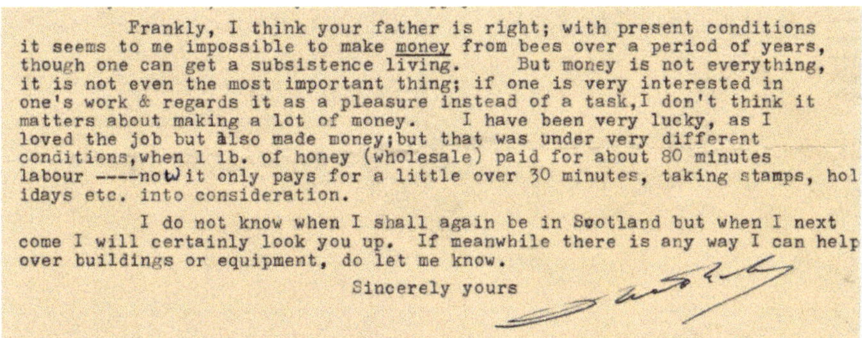

Frankly, I think your father is right; with present conditions it seems to me impossible to make money from bees over a period of years, though one can get a subsistence living. But money is not everything, it is not even the most important thing; if one is very interested in one's work & regards it as a pleasure instead of a task, I don't think it matters about making a lot of money. I have been very lucky, as I loved the job but also made money; but that was under very different conditions, when 1 lb. of honey (wholesale) paid for about 80 minutes labour ----now it only pays for a little over 30 minutes, taking stamps, holidays etc. into consideration.

I do not know when I shall again be in Scotland but when I next come I will certainly look you up. If meanwhile there is any way I can help over buildings or equipment, do let me know.

Sincerely yours

A.W. Gale, 31st October.

1959

Achalader, Blairgowrie

Production figures refer to honey actually produced. They have little to do with sales, and do not include any honey purchased elsewhere. They do include any seed honey added.

The weight of honey being taken back to the farm from the heather caused all the springs in the Landrover to break!

Crop: 38,700 lb or 17 Tons 5 Cwts 60lb

Achalader – February.

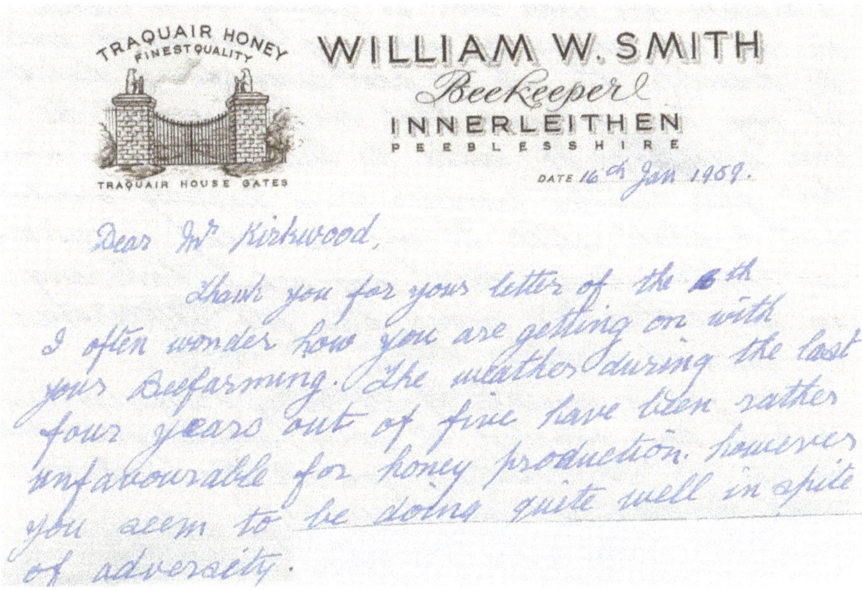

TRAQUAIR HONEY
FINEST QUALITY
TRAQUAIR HOUSE GATES

WILLIAM W. SMITH
Beekeeper
INNERLEITHEN
PEEBLESSHIRE

DATE 16th Jan 1959.

Dear Mr Kirkwood,

Thank you for your letter of the [illegible]th I often wonder how you are getting on with your Beefarming. The weather during the last four years out of five have been rather unfavourable for honey production. however you seem to be doing quite well in spite of adversity.

William W Smith, 16th January.

Busman's holiday to Norway.

1960

Achalader, Blairgowrie

Production figures refer to honey actually produced. They do not include any seed honey added, or any honey purchased. As clover is added to the heather honey before sale, these figures will not coincide with sales figures.

Crop: 41,754 lb or 18 Tons 12 Cwts 90 lbs

Fiona (the bee-daughter) was born

1961

13,560 lb or 6 Tons 1 Cwt 48lb

Bob Barrie (Foreman) and me – Mount Blair.

March HPA – Seale-Hayne College, Newton Abbot – Norman Stewart, me and George Wallace of South Molton.

With R.O.B. Manley, Chiltern Honey Farm, Preston Crowmarsh, Benson, Oxford.

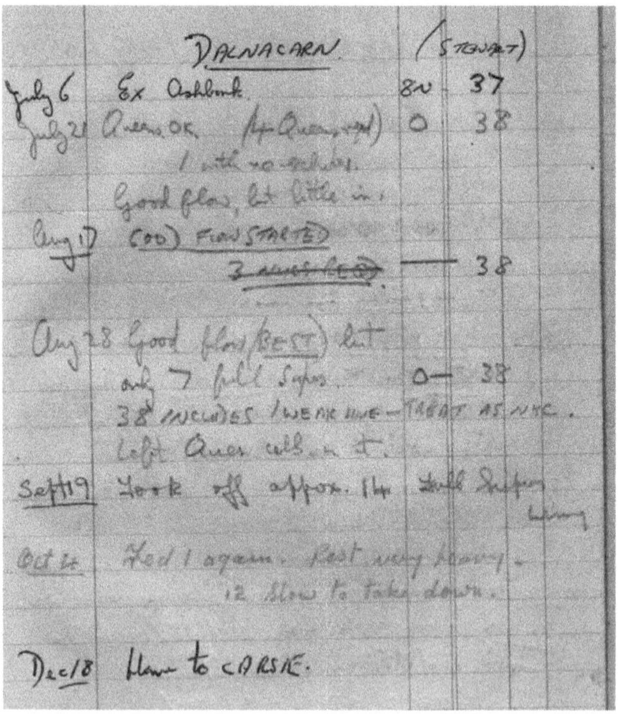

DALNACARN (STEWART)

July 6 Ex Ashbank 8N 37

July 21 Queens OK 0 38

Good flow, but little in.

Aug 17 (OD) FLOW STARTED

— 38

Aug 28 Good flow (BEST) but only 7 full supers. 0— 38

38 INCLUDES 1 WEAK HIVE — TREAT AS N.C.

Left Queen cells in it.

Sept 19 Took off approx. 14 full supers

Oct 1st Fed 1 again. Rest very heavy. 12 Slow to take down.

Dec 18 Home to CARSIE.

Dalnacarn record.

PRESTON CROWMARSH,
BENSON,
OXFORD.
WALLINGFORD 3165.

June 20th 1961

Dear Mr Kirkwood,

Many thanks for the two photos which I am told are good, but I had no idea my hair was that white. How little we know ourselves. Good of you two as well.

You have had a better time than we, I should think. We have used 18 tons of sugar this time, which is heavy for us after heavy winter feeding; but there was no incoming nectar at all until about five days ago. Now bees are collecting quite a bit. We had a splendid rain all day one day last week which did the trick. I was getting alarmed in case of a continued drought. However, we will soon be wanting more rain and it does not look like it at all It is cold at night. Since charlock and sainfoin d departed this life, we get practically nothing before July. We may have a good season or a bad one -- only time will tell; so we must possess our souls in patience and meanwhile try to scratch part of a living from our imported trade.

Planning a honey farm. That depends on the size of the propoed business, but unless you can do the job on a fairly large scale and employ a good deal of labour, its a poor look-out. Its the scale that does it.

I don't know if I shall ever get around to another visit to Scotland, but I will pay you a visit if I do and many thanks for the invitation. But I am too deaf now to be much good to anyone.

Yours sincerely

Robt Manley

R.O.B. Manley, 10th June 1961.

Trip to Buckfast Abbey.

Brother Adam's hives.

Brother Adam being interviewed.

Note

For further information on Brother Adam (Karl Kehrle OSB OBE), (3 August 1898 – 1 September 1996) please refer to the Bibliography.

Pristine extracting room.

The incubator.

Buckfast Heather Honey.

Many thanks for your letter & we will send the honey as soon as we can get a through truck. I think you will find it some very nice stuff; it was exceptionally good quality this year as well as being a heavy crop. We finished up with 88 tons 14 cwts.

You certainly have gone ahead with the bees! I must try to visit you some day when I am in Scotland; I have been there twice this year, once with the British Grassland Conference at Aberdeen, another time staying with friends in Kincardineshire & again near Aberdeen. There is just one little point I would like to make; probably the advice is not needed, but as you know, I have been in this business for well over 40 years & have seen quite a lot of things happen to quite a lot of folk during that time. Don't get more colonies than you can handle profitably; it pays better to look after a smaller number & run them well than to get a very large number which cannot be given proper attention at the proper time.

With all good wishes

Sincerely yours

P.S. Being collected by Railway 16/11/61.

A.W. Gale, 15th November 1961.

Ian – the future bee farmer, was born.

1962

Glenardle, Bridge of Cally

Bill Lisle, Finegand Apiary. I had several VW T2s – workhorse of the 1960s.

Moved to Glenardle at Bridge of Cally.

Finegand honey.

Cordingley Apiary, Shropshire – Me, Mrs W.W. Smith, Willie Smith, Norman Stewart, David Spence and William Selby Robson.

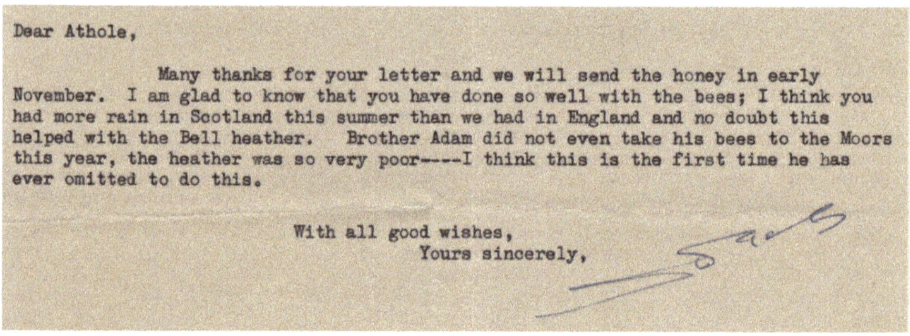

Dear Athole,

Many thanks for your letter and we will send the honey in early November. I am glad to know that you have done so well with the bees; I think you had more rain in Scotland this summer than we had in England and no doubt this helped with the Bell heather. Brother Adam did not even take his bees to the Moors this year, the heather was so very poor----I think this is the first time he has ever omitted to do this.

With all good wishes,
Yours sincerely,

A.W. Gale, 9 October.

1963

Glenardle, Bridge of Cally

Production figures refer to honey actually produced on the farm. They do not include any seed honey added, or honey purchased. As clover is added to the heather before sale, these figures do not coincide with sales figures. (In actual fact all honey was sold as heather this year).

This season was very very poor, with heavy winter losses to start. Those colonies left alive were very weak. The raspberries yielded moderately well, but colonies only built up on this. The bell heather and clover failed, apart from a few days of heat when clover yielded. The rain spoilt the best of the ling, but there was a good flow for a few days at the end of August. If we had not fed sugar in August the season would have been a disaster.

International relations all well six months after the Cuban missile crisis. Myself, Nikolai Mikhailovich Glushkov, director of the National Science Institute of Beekeeping in Rybnoye, Russia and Norman Stewart at the HFA – Swaffham, Norfolk.

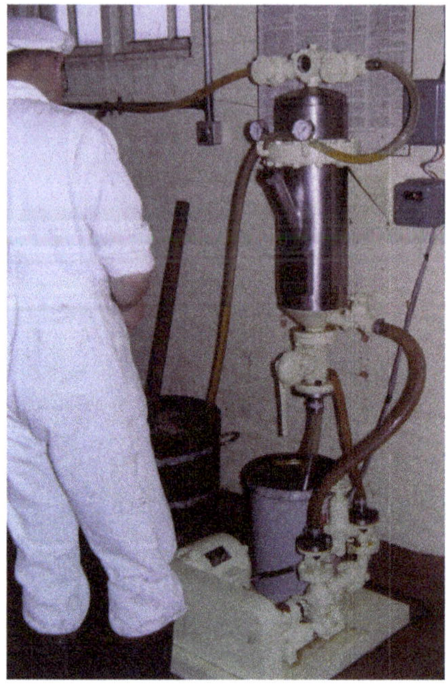

Visit to Cecil Tonsley's Honey Farm, Coton Lodge. Bottling, filtration.

Visit to E.W.D. Madoc's Honey Farm in Watton, Norfolk.
Norman Stewart, Teddie Madoc and myself.

Note

For further information on Edward Warren Darby Madoc-Jones (1900–1982) please refer to the Mattishall village website in the Bibliography.

Teddie Madoc has 1800 hives arranged in eights in the apiary to take up less space. Six apiaries are checked per day. Only half of them are examined carefully if the other half are okay. This is done by starting at the opposite end of the apiary each time. British Standard frames are used and 12 cell cups are put on each of 2 bars in 1 frame. The cups are staggered on the bar to spread them out more. Grafting is started on 1st May and continues till all the queens have been raised. They say that they use a proportion of BS 11 frame hives and MD supers and find there is no difference in the crop.

Only 500 colonies are moved to the fruit orchards. The pollination fee is critical to the business.

Catriona is born.

1964

Production figures refer to honey actually produced on the farm. They do not include seed honey added, or honey purchased. The Bell Heather figure should be higher, and the Ling a bit lower, but this is very difficult to assess.

This season started fairly well, but spraying caused serious weakening of the colonies working the raspberries. The Clover flow was poor in most areas, but there was a very heavy Raspberry flow where spraying did not occur. The Bell Heather was very good, and some apiaries made over 112 lbs on this one crop or about 150 lbs altogether.

The first of the Ling was spoilt by rain, but a good flow came on later from some apiaries.

Clover – Culthill, Caputh.

Smith combs being filled with heather.

Willowherb, Achalader.

Bill Lisle and Harry Kean, Cluniemore.

Learning the ropes!

Norman Stewart and David Rowse, Clapper Hill Honey Farm.

Mrs Rowse bottling.

Note. For further information on David Rowse please refer to the Bibliography.

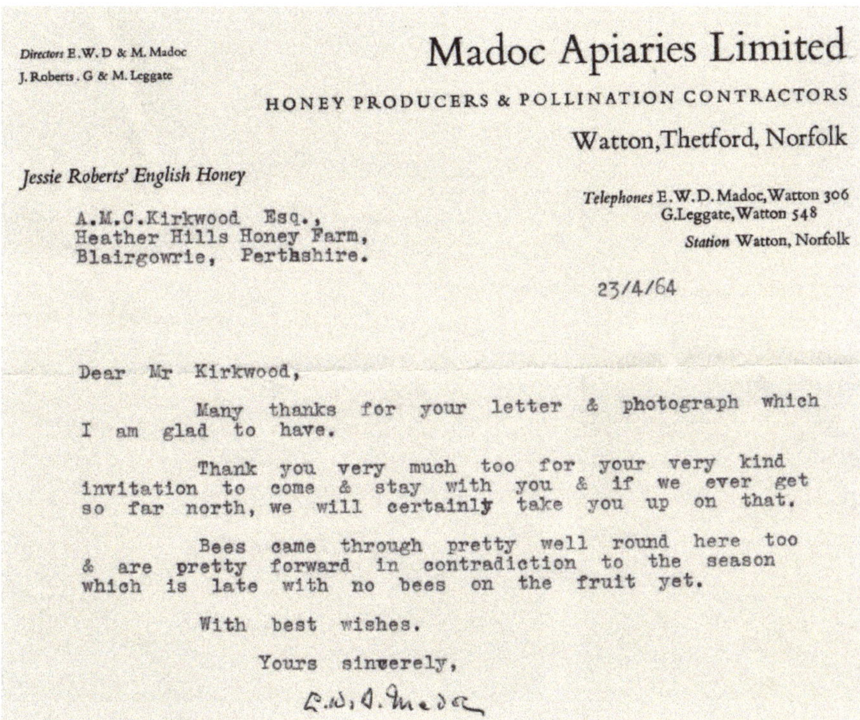

Directors E.W.D & M. Madoc
J. Roberts. G & M. Leggate

Madoc Apiaries Limited

HONEY PRODUCERS & POLLINATION CONTRACTORS

Watton, Thetford, Norfolk

Jessie Roberts' English Honey

Telephones E.W.D. Madoc, Watton 306
G. Leggate, Watton 548
Station Watton, Norfolk

A.M.C.Kirkwood Esq.,
Heather Hills Honey Farm,
Blairgowrie, Perthshire.

23/4/64

Dear Mr Kirkwood,

Many thanks for your letter & photograph which I am glad to have.

Thank you very much too for your very kind invitation to come & stay with you & if we ever get so far north, we will certainly take you up on that.

Bees came through pretty well round here too & are pretty forward in contradiction to the season which is late with no bees on the fruit yet.

With best wishes.

Yours sincerely,

E.W.D. Madoc

Teddie Madoc, April 1964.

1965

Production figures refer to honey actually produced on the farm. They do not include seed honey added, or honey purchased. Bell and Ling were very difficult to estimate in proportion, but the total figure for the two is correct, and there was about the same tonnage of each.

The season was very bad, and the size of the crop was very surprising.

Harry Kean checking Smiths at Pitcarmick.

Transferring Smiths to jumbo.

Achalader Honey Farm from 1954-1966.

John Manley, Chiltern Honey Farm, Kelley bottling machine.

Drumellie Apiary.

Trip to France with Francis Cordingley. M. Michaud, Barbara and Francis.

M. Michaud's new premises, Pay, Pyrenees.

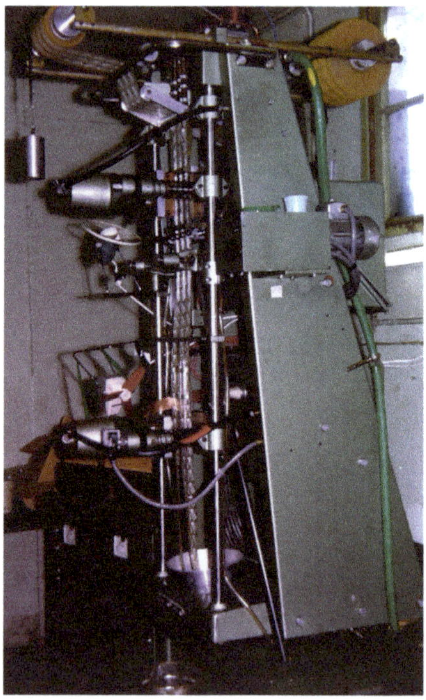

Plastic bag machine.

1kg pots.

M. Thomas' house, Fay-aux-Loges.

Francis Cordingley and M. Gilet, Chateau Apiary.

With M. Thomas, Fay-aux-Loges.

1966

Last year at Achalader

Production figures refer to honey actually produced on the farm. They do not include seed honey added, or honey purchased elsewhere. It does of course include any syrup which finds its way into the supers. In 1966 the bell crop was very poor, and feeding was commenced in late July to avert disaster. The ling started to yield heavily however, and 18 tons were extracted. Another 2 tons of syrup or clover were extracted with the loosener at the same time. Extracting was finished on November 2nd.

In the winter preceding this crop losses in Smith hives were heavy, and colonies which lived through the winter were weak in most cases. The 9-frame Dadant hives however were mostly in good condition in Spring, and practically none died out. Some Smith hives, with no feeding at all produced over 100 lbs of Ling honey.

Comb honey.

Ling heather.

BFA Pembroke College, Cambridge. My rooms!

The Howatsons, Garlieston, Newton Stewart.

Howatson double nucleus.

Good Smith super of ling.

At the summit of Goat Fell, Arran.

Always on the lookout for bee boles (Arran).

Sarah was born in 1966.

1967

Heather Hills Honey Farm, Bridge of Cally

1967 was a very unusual season, with results similar to the previous season. June was too dry for raspberries to yield, and most of the early crop was from honeydew which greatly darkened all the flower honey. There was however a flow from one or two raspberry sites.

Bell heather was fairly good and after about July 24th, the season was absolutely perfect with hardly a day lost till the beginning of September. Supers were filled very quickly, and, in due course, we ran out of them, and had to extract and return very quickly. Glenshee Lodge was perhaps the best apiary with 100 lbs long on every hive.

Balnakeilly – very good ling crop on Smiths.

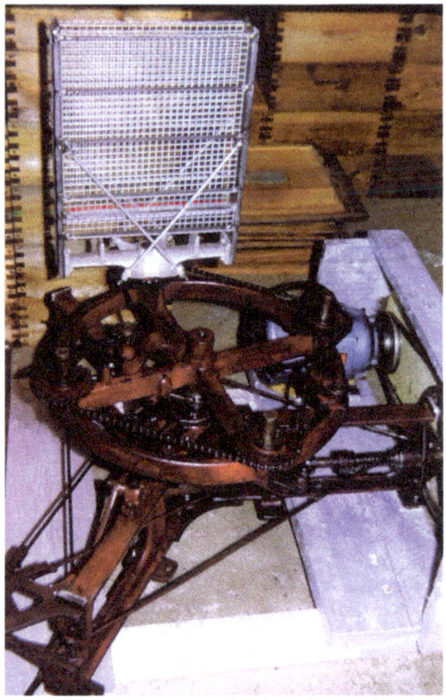

Lewis-Markle (1924) extractor with and without drum. R.O.B. Manley's original machine.

Superior Lifetime extractor (new).

Sherpa stacker handling full ling supers.

Big load of supers being moved from Achalader to the new honey farm at Bridge of Cally.

Holgates Honey Farm, Aberayron, Cardiganshire visited by myself and Norman.

Thomas French loosener – 22 tons ling!

Fonab Forest – very good crop.

Valerie was born in 1967.

1968

1968 would have been a bumper season, but June was too dry for a good flow, and July was unsettled. The bell heather failed to yield any reasonable quantity. The best flow in our own apiaries was in Glenerrochty where there was a heavy flow at bell heather time. Source unknown. Fonab was also excellent. Ling very good to West.

Clover yielded very heavily in the West of Scotland and at Aberfeldy it was the best heather season for a long time. Bees East of Cally were rather poor.

I decided that as most of the machinery I would require for a large scale bee farm was from abroad, it would be sensible to begin to change to Langstroth hives. 20 were ordered this year.

Smith feeders on Langstroth hives, Pitcarmick Apiary.

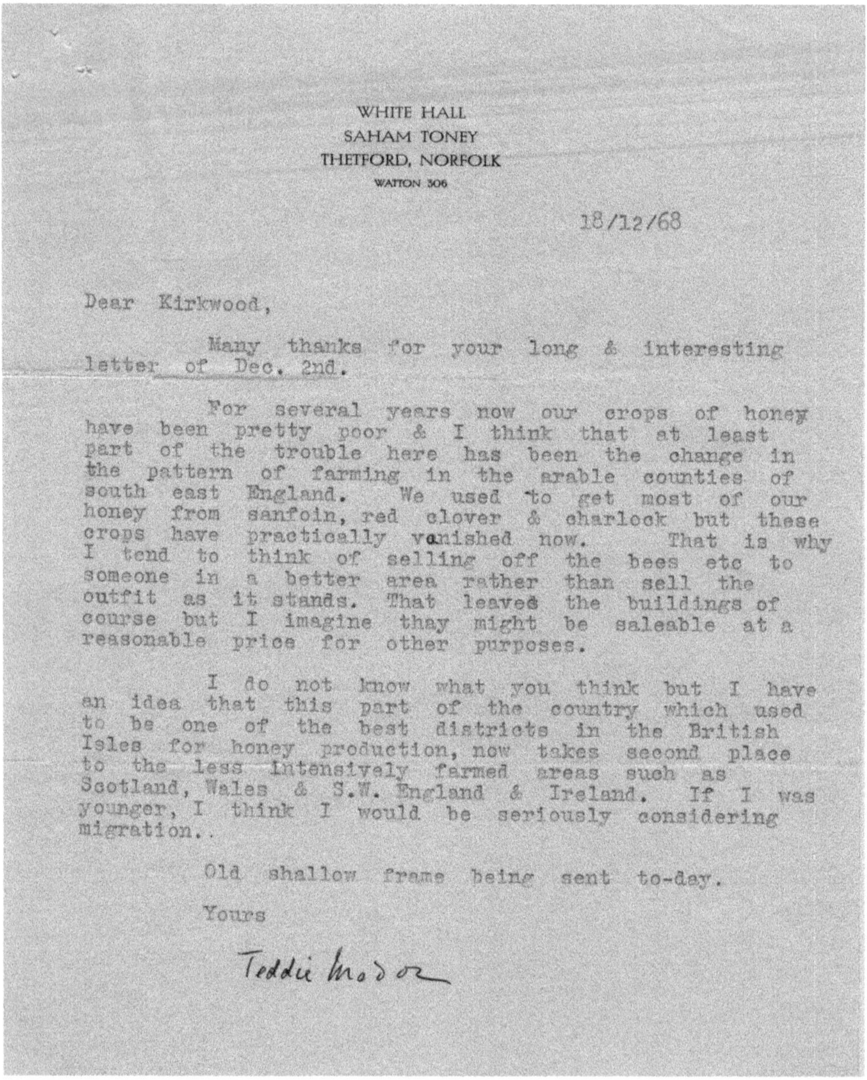

WHITE HALL
SAHAM TONEY
THETFORD, NORFOLK
WATTON 306

18/12/68

Dear Kirkwood,

Many thanks for your long & interesting letter of Dec. 2nd.

For several years now our crops of honey have been pretty poor & I think that at least part of the trouble here has been the change in the pattern of farming in the arable counties of south east England. We used to get most of our honey from sanfoin, red clover & charlock but these crops have practically vanished now. That is why I tend to think of selling off the bees etc to someone in a better area rather than sell the outfit as it stands. That leaves the buildings of course but I imagine thay might be saleable at a reasonable price for other purposes.

I do not know what you think but I have an idea that this part of the country which used to be one of the best districts in the British Isles for honey production, now takes second place to the less intensively farmed areas such as Scotland, Wales & S.W. England & Ireland. If I was younger, I think I would be seriously considering migration..

Old shallow frame being sent to-day.

Yours

Teddie Madoc

Teddie Madoc, December 1968.

Steam raiser uncapper.

Lifetime extractor and radial.

Visit to Auchincruive.

Stewarton hive, Ian Maxwell, Auchincruive.

The Stewarton hive.

Bee boles, Cornwall – Mr Philips and Norman Stewart.

Visit to Bernard Ardill's, Dalditch Apiary, Budleigh Salterton, Devon.

Bernard Ardill and Norman Stewart, Dalditch Apiary, Budleigh Salterton, Devon.

Tom Bradford with skep.

Tom Bradford's extracting room.

Editor: Mum and Dad took the three older children on our first foreign holiday, to Marjorca, the "babies" being left in the care of our grandmother. We have wonderful memories of quiet beaches, delicious breakfast ensaïmadas and playing ping pong under the palm trees. Unfortunately, however, this was also the year when mum discovered that she was allergic to bees! She was swimming in the sea when she was stung. All we recall, as children, is that we were left on the beach to be looked after by a kindly tourist while our father and the hotel staff dealt with the dreadful panic of the situation. She has since had treatment and carries an Epipen but has never been stung again.

All the beehives which used to be wintered around Glenardle were moved away, from then on Dad removed his bee suits and clothing and put them straight in the washing machine himself so that mum didn't have to be in contact with them. A theory, new at the time, was that the families of beekeepers were being exposed to their clothing but never actually being stung and were building up a possible allergic response. It has been found now that many wives and children of beekeepers are allergic to bee stings, perhaps one third of BFA member wives being allergic but not realising at the time that they were not alone.

Norman Stewart, Lint Mill, Inverar, Glen Lyon.

1969

1969 was a wonderful summer apart from a cold spell in mid-August, and a wet fortnight in July. Very good yields were obtained from the raspberries in the hot humid weather which lasted much of the summer.

Clover did not yield much at all however and the bell was largely a failure. The bell type honey obtained may have been from heath or trees, but little came from the real bell heather.

Ling was frosted very badly on all lower hills, and good crops were only obtained on higher ground, and in a few areas. The yield was therefore surprising, and only accounted for by the increase in colonies, and much feeding in some apiaries.

The total of Langstroth hives now amounted to 300.

Fonab Forest (Langstroths with Smith nucs).

Langstroth comb of ling.

Langstroth NZ – the bee clearer board..

2800 Langstroth frames just wired.

Blocks of wax from the melter.

Rietsche embossing mill –
7,000 sheets made in 1969.

Norman with the Robsons'
dolls-house beehive.

Norman with Willie Robson, Mrs W.S. Robson, Mrs Pat Bradford, Tom Bradford and William Selby Robson, beekeeping advisor, during the Bee Farmers Association conference, North Berwick.

Visiting Les Thorne at his wax plant.

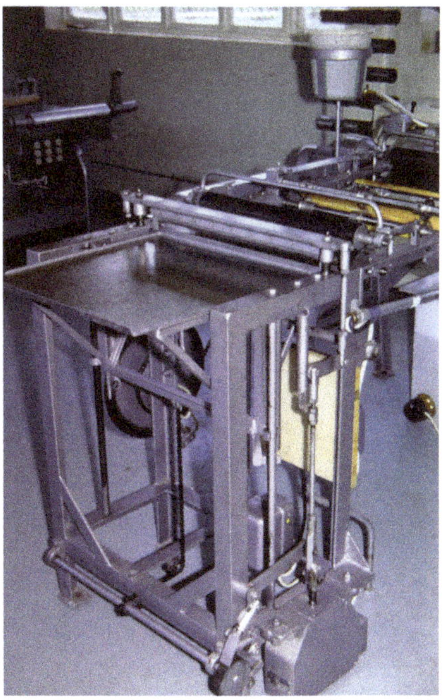

Thorne's wax foundation mill.

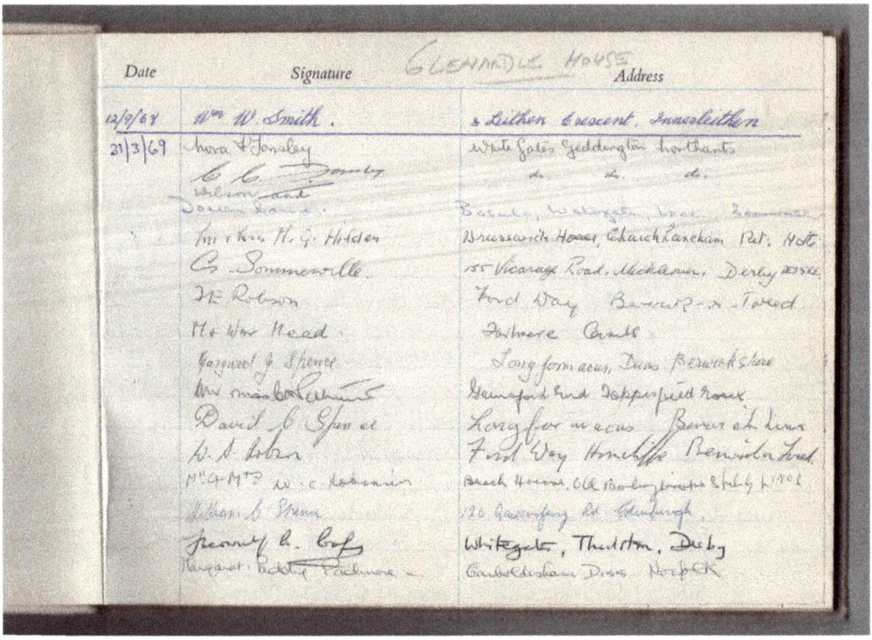

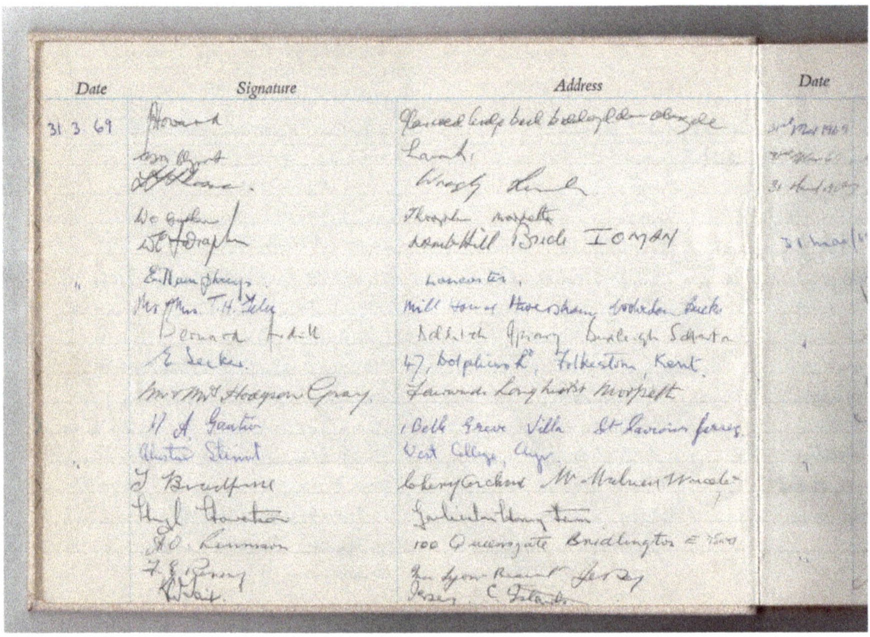

Heather Hills Honey Farm Visitors Book 1967-69.

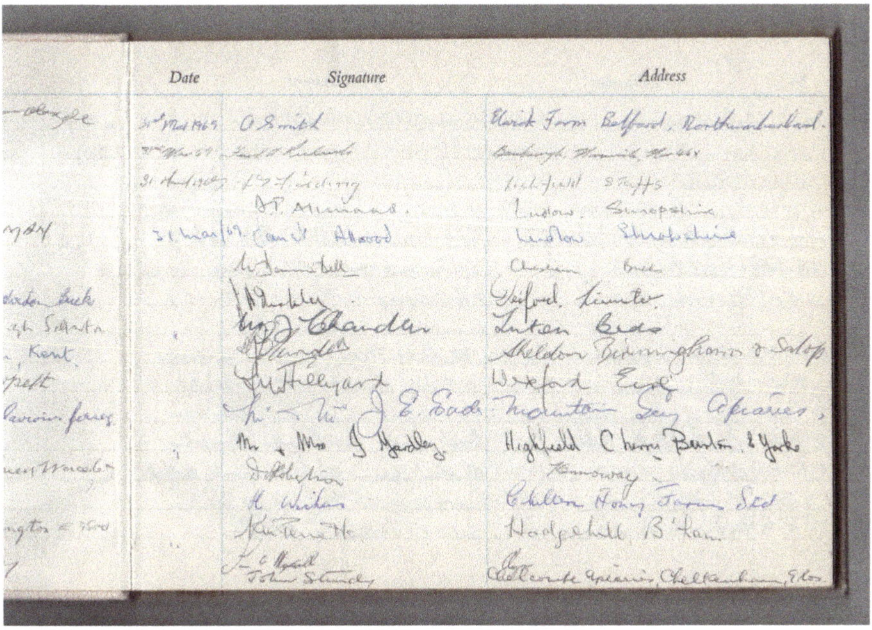

In 1967 Dad had acquired a Visitors Book and asked Willie Smith to be the first to sign. It was then left aside until 1969 when an Open Day was held at the Honey Farm and, as can be seen here, it was a successful event with, among many others, Cecil and Nora Tonsley, William Selby Robson, Tom Bradford and Les Thorne all in attendance.

J.E. Eade and Norman Stewart at Mountain Grey Apiaries.

1970

1970 was a very bad Spring for the bees and losses were heavy. Spring came a few weeks too late. Only about 500 colonies wintered in good condition. The rest were weak and many were dead. All over wintered nuclei were lost. 437 shook swarms were required to restock.

June was very hot, and enabled increase to be made under good conditions. July was very wet. Bell would have yielded well, but the weather did not give it a chance.

The first fortnight of August was very good. Unfortunately, a lot of the heather was poor, and so yields were very erratic. There was a very good crop at Tummelbridge. The Angus glens were rather poor though good on high ground. The yield was increased greatly by heavy feeding of half the hives.

Heather Hills Honey Farm. Photo: Willie Henderson.

Bob Barrie, Foreman and Harry Kean milling a plain beeswax sheet.

Sid Fender producing long embossed sheets.

6kw Boiler.

Two 300 gallon honey tanks.

Syrup mixer.

Ian at Straloch.

Thoughts on Langstroth Management

Use one double nucleus (standard broodbox) to every 8 hives or so. Double nucleus to consist of a broodchamber with removable division board in the centre. A divided floor and twin inner cover. Also standard roof. 300 or even 250 of these broodboxes would do for a bee farm of 2000 hives.

At the end of May or early June make as many hives as possible up to 9 frames of brood and remove the queen and 1 brood to a nucleus. Because the nuclei are used only for storage of queens and not for mating the double size will present no problems. A lot of hives will be left with only 3 or 4 frames of brood and a queen, so they will give little or no trouble with swarming. They should be productive on the heather. Those dequeened hives with 8 or 9 frames of brood are left with a queen cell after 9 days. This means 7 day working for about 10 days at the beginning of June.

Inspections after this are carried out only every 14 days, so it is possible to work a 5-day week unless the weather sets back the work very considerably. A 6-day week would be worked during July to get the raspberry honey off, and the bees moved in the quickest possible time.

At the first inspection at the heather, hives would be either queenless or with a young mated queen or possibly building up well (if they had been robbed earlier of brood). The queenright hives would be alright for the rest of the year. The queenless ones would get a nucleus united back with the queen caged and the stronger ones building up could be interchanged with weaker hives if they were building cells for swarming.

Nuclei could be advertised for sale in July, and if not sold, could be doubled up for wintering by removing the division boards. They would be fed with standard Miller feeders. (Note: make these feeders with a division up the centre of the access space to keep bees apart). There would be no problem with extracting brood frames if this was not desired.

Due to the fact that dequeened hives in early June would need about 9 frames of brood this would mean that not every hive could be dequeened. Many would be left rather weak, so they would give no trouble until heather time. This means that not a great number of nucleus hives would actually be required.

In the event of the system being changed there would be no expensive nucleus hives to scrap.

The twin inner cover could be made by using an ordinary inner cover with a cloth cover below, the cloth cover nailed up the centre to the top of the division board. (or it might be possible to saw up Smith inner covers for this purpose).

The floor would be an ordinary floor capable of being made back into a standard floor by replacing a few parts.

Dead hives in the spring would be refilled first of all from any over-wintered nuclei. Others with artificial swarms.

5-frame nucs at Kinloch.

Propolis seal on Norman's hive at Glen Lyon (roof knocked off).

Langstroths wintering at Fonab camp.

1971

Winter losses were again rather heavy, especially in some apiaries where queens had not mated well the previous Autumn. June was too dry at first and then too cold and wet for a good raspberry flow. Clover yielded very well at the beginning of July around Bankfoot and Methven, but not around Blairgowrie. This was the first year that bees were moved into Aberdeenshire, and results there were very good indeed. Some hives at Ballintuim, by Braemar had 4 supers of bell and 5 supers ling – about 260lbs in all.

Bell yielded quite well in Perthshire too, and well at the top of the Angus glens.

Ling yielded well everywhere, which is very unusual. Extraction of ling started Sept 4th and finished October 26th. (Extracting was stopped for one week during this period.)

Left to right – Charlie Macdonald, John Donaldson, Willie Henderson, Harry Kean, Stanley Bednarek, Carol Upton and Bob Barrie (Foreman).
Photo used by kind permission of D.C. Thomson & Co. Ltd.

BFA Spring Meeting, Malvern. Photo attributed to Terry S. Lowe, Retford.

Visit to Bulmers Cider, Padmore.

Willowherb.

French loosener (Thomas) with steam raiser and hydro extractor in background
Photo used by kind permission of D.C. Thomson & Co. Ltd.

Excerpt from *Lincolnshire Beekeepers' Association Year Book*

Honey Looseners
by A.M.C. Kirkwood

In certain parts of the world a thixotrophic (jelly-like) type of honey is produced – notably Ling heather *Calluna vulgaris* in parts of Northern Europe and Manuka (Tea Tree) *Leptospermum scoparium* in New Zealand.

Until the invention of the honey loosener these honeys were usually eaten in the form of honeycomb or pressed. In both cases the loss of drawn comb reduced subsequent honey crops.

... Even after loosening, however, Ling honey is still very thick and it will not readily extract in a radial machine, no matter how fast it is driven. A tangential machine is required and, for large scale work an automatic-reversing type is very desirable.

... Thixotrophic honey in cappings is difficult to deal with, as it will not separate out from the wax in a capping smelter, but simply burns. It is necessary to use pressure. For this the honey press can be used, but a centrifuge makes a better job, and gets the wax really dry. The speed of the centrifuge must be high. A 14" hydro-extractor running at 1,450 revs/minute is suitable.

Thixotrophic honey means a lot of extra work, but the higher price received for the end product makes it all worthwhile.

The Solheim Norwegian honey loosener.

Bottling. Photo used by kind permission of D.C. Thomson & Co. Ltd.

Honey sump for the extractors features a float-operated pump control.

Catriona and 25 gallon drums of ling honey. Photo: D.C. Thomson.

Yale lift truck and Langstroth pallets.

16-frame Superior Lifetime extractor, Stewart Honey Co, USA.

At the Fonab Forest apiary.

1972

We held an Open Day at the farm on the 24th June with visitors from all over – America, South Africa, Ireland, Norway and Belgium.

The season of 1972 was very bad until mid July. There was a great deal of rain and cold weather. When the heat wave came in July it reached 84°F. for several days, but it was too hot for a good flow. The bell heather was very poor indeed, and only reasonable in Aberdeenshire.

A good ling flow started on August 19th and continued for about three weeks with only a poor day here and there. Feeding heavily had been commenced and this programme was finished about the end of the third week in July. The crop would have been a good bit lower without the feeding. Extracting finished on November 7th, later than in the previous year, although there was much less ling honey.

256 colonies of bees were sold in Smith hives from October 1972 – April 1973. Richard and Joyce Wilson of Cych Valley Honey Farm in Wales bought a mixture of Smith and older Langstroth equipment. In spring they move their colonies to Kent for contract pollination.

BBC Scotland 'Farm Forum' Team – Pat Chalmers, Balnakeilly Apiary (August). No-one was stung!

Hives used as advert at Glenardle.

Winch VW.

Langstroths at Rhidorroch Apiary.

Dell Dalcapon.

Norman, Fonab Forest.

Rex Miller with Norman Stewart.

1973

This year the Bell heather was very promising indeed and started to yield heavily in most of the traditional Bell areas. However, rain came down heavily about 20th July and this finished it off. The bloom on the ling was the best colour for many years, and the weather became hot during August. However, this dried up the heather somewhat, and yield might have been better if there had been a little rain. There was also a very considerable loss of ling honey due to granulation.

140 colonies were sold in Autumn 1973.

Because a new pump did not arrive in time ling extracting did not start until October 27th. The 3000 supers were extracted by November 23rd.

Cornwall – Mr and Mrs Bray and son Richard with Norman.

Norman at BFA Eastbourne.

Wild orchids.

1974

Production figures refer to honey produced on the farm. They do not include any honey bought in. The figure may, and in this season certainly will, include quite a lot of syrup stored in the supers.

The season started well with a good flow from raspberries in Dundee, Alyth areas and some parts of Blairgowrie. Other districts were poor. There was very little spray damage and good colonies stored 60 lbs.

The weather broke down at the end of June after a very good spell, and it became wet and very cold. Colonies moved early to the hills suffered badly. They would have done a little better on the low ground. In the hills it remained cold all summer.

The July crop contained quite a bit of syrup, but also included some honeydew and lime. Bell would have yielded well if the weather had been drier.

The weather in Perthshire and Angus was very poor, with very little bloom. In Aberdeenshire it was fortunately good, and most of the ling honey came from the 12 or so apiaries up there.

16 tons of sugar were fed during July and August and early September. The sugar was bought at £120 per ton before the price began to rise.

Bees on knapweed, Tummelbridge.

Doreen and Wilson Carne, bee breeder, with Catriona, Polzeath.

Barbara with Pat and Brian Sherriff (of B.J. Sherriff Clothing), Cornwall.

1975

Production figures refer to honey actually produced. They do not include any honey bought in. Sugar was fed this year only to the 700 artificial swarms made up in June, and to some of the colonies in September. There would be no syrup in the supers this year.

There were heavy casualties in early Spring resulting from the previous bad summer. Many colonies which overwintered died later when the queens became drone-layers. This accounted for most of the losses.

The weather turned very not after mid-July and continued in the 70s and 80s right to the end of August with few breaks. The raspberry flow was reasonably good, but clover, Bell heather and Ling were all superb. Especially the Ling.

A group of German visitors came to see the farm in August. In September Alan Bennett from Pender Bros, New South Wales and Kevin Ecroyd from New Zealand visited.

Bearing in mind that a large number of the supers were new and contained half foundation the result was very good indeed.

The Ling crop was extracted quickly, finishing on October 9th. We packed 34 tonnes in 40-gallon drums for export to Germany.

The last of the 1050 Smith hives were sold in December.

Oil seed rape, Rosemount.

1976

Spring losses of colonies were very heavy at 25%, mainly due to lack of attention the previous autumn. Frames of brood should have been given earlier to queenless hives so that queens could be raised, but this was not done until too late in very many cases.

The raspberry flow started well but stopped after a week. There was no flow after that in this area until August 14th when it rained for one day. There was heavy dew all summer, but it was insufficient to give a honey flow. It was the longest and hottest summer in living memory. For about 6 weeks it was hot sunshine all the time, and temperatures were in the 85-90°F.

Bell heather was in full bloom but did not yield except for a day or two, and there was none to extract.

Feeding had to be carried out late July – about a gallon being fed to most colonies.

It was decided to experiment with moving the bees from Scotland to the orchards of Kent to benefit from the pollination fees of the fruit orchards there. One hazard that was unforeseen was that the drivers found themselves carrying a load of bee hives down The Mall in the middle of the Queen's Silver Jubilee celebrations.

Photos used by kind permission of D.C. Thomson & Co. Ltd.

Aro 4x4 van, Tomintoul.

Glenisla – visit by the Carnes and the Brays.

Dalvey apiary.

Ian with me on a fly fishing trip.

1977

Lecture by Athole Kirkwood from a conference on 26 January 1977 Opportunities for Commercial Bee-keeping in the UK National Agricultural Centre, Kenilworth, Warwickshire

The title of this conference is "Opportunities for commercial Bee-keeping in the United Kingdom". My own experience has mainly been confined to the East and West of Scotland, apart from a brief spell with Mr A.W. Gale in Wiltshire.

Scotland can be divided into the East and West.

The West is like another country, being much milder. Some small areas on the coast get great benefits from the Gulf Stream. The area tends to be rather wet for full-time bee-keeping, but it is practiced, and in good summers the clover yields very heavily. The heather too can give good crops. The mild winters tend to encourage the development of Acarine disease, but apart from this wintering is fairly easy.

The East, where my own farm is situated, contains a lot of very good agricultural land, and not so much rough pasture, except in certain areas, so the production of clover honey is limited. However, we do

New farm building at Bridge of Cally.

grow 80% of British raspberries in Perthshire and Angus, and big yields of honey were obtained from this plant at one time. Spraying, however, is a very serious problem today, and good crops are not frequent. Heavy losses of bees occur from time to time.

A lot of the best heather districts are situated in the Eastern area, and because of the drier climate the flow tends to be more reliable. But bad seasons do come. The rain sometimes forgets to stop.

I started beekeeping 31 years ago with a couple of WBC hives, and gradually built up to 50 in Smith equipment. In 1955 I took the plunge and went full-time. I only had 200 hives, but always having been lucky, had a fantastic year and finished up with a very good crop.

Today, we have just over 2000 colonies, and during the last seven years have standardised completely with new factory-made Langstroth equipment.

The farm itself consists of two Atcost-type buildings with double-wall construction and a total floor area of 7000 sq ft. In addition, there is a large loft with an area of 2400 sq ft. Incorporated in the layout are a warm room, an extracting room, a bottling room, a workshop and a cool store. The loft will hold 30 tons of sugar if necessary, and all the cartons that we can use in a year. It is also used for the storage of empty drums. We have drum capacity for 120 tons of honey.

These buildings were put up within the last 10 years, but at today's prices would cost about £30,000. Grants were obtained from the Department of Agriculture for one building and from the Board of Trade for the other. Prior to the 1971 Agriculture Act bees were not officially classified as agriculture, and so we sought and obtained a grant from the Board of Trade for our first new building. The buildings are de-rated, so no rates are payable at present.

The equipment inside the buildings, excluding vehicles, has cost about another £30,000, and here again Development Grants were obtained. Usually, the rate was 20%.

The hives represent the greatest amount of capital of any bee business. Fortunately for us, these were mostly purchased before the recent steep escalation in prices, but even so they cost a great deal of money, and no grants were obtainable. The bees, however, paid for the hives themselves, which is the safest way to buy them, the whole honey farm having been built up from an initial borrowed capital of £2,000.

I believe it is a great advantage to start any business in a small way, and to build up gradually out of profits. In other words, to start at the bottom. This applies particularly to bee farming, where it is very easy to lose a lot of money quickly. I have not the slightest doubt that if I had started with a lot of capital behind me, I would no longer be a full-time beekeeper, but would be working for my living elsewhere. The money would have gone!

However, there is no point in my telling the keen young beekeeper that it is impossible to make a living from bees today. This is exactly what I was told by the experts when I started, and I am afraid that it simply made me more determined than ever to succeed.

It is very important to remember in the beginning that what you are after is honey, and that only the bees can produce it. So every penny that can be spared should be guided in that direction. Buildings, machinery and plant are of secondary importance to this. They help with the work certainly, but they produce nothing. They should only be bought out of profits. Buildings, in particular, can use up a great deal of capital, and at the outset should always be rented if at all possible. Nothing complicated is required. I built up to 1000 hives while renting some old stables. When we went there, there was no electricity and only a cold water supply. I would recommend that you get electricity at an early date however. It is no fun to extract 10 tons by hand!

The hives being the most expensive part of the whole operation, it is very desirable to make the right choice at the beginning.

I was quite convinced back in 1947 that the Langstroth had the most advantages, but I was talked into adopting the Smith type, and did so for the next 20 years, with no disastrous results! They are excellent hives for the man who wants to run up to about 500 on his own. But for large-scale use they proved to be a great nuisance, and whenever I could afford it we made the big change back to Langstroths. This took seven years, but was well worth the expense and the trouble involved. It is the hive I recommend to anybody starting serious bee farming. After all, it is used to produce most of the surplus honey in the modern world and any new innovations that come in from Canada, the United States, France, Australia or New Zealand will be based on the Langstroth dimensions.

Already we have from Kelleys of Kentucky the complete polyethylene hive outer parts. I would not recommend it here without a lot of field trials, but it points the way in which progress will be made.

I would like to see a big swing towards the adoption of the Langstroth hive as a standard in this country too, but with prices as they are I think we would need a Government or EEC subsidy to get things moving!

In the supers we used the Modified Dadant frame of Manley pattern. The frames are made slightly wider than recommended by Mr Manley so that eight fit nicely into the super. I was interested to hear recently from some visiting Australians that exactly the same frame has been adopted in their country. It has proved to be the best possible type for the latest automatic uncapping machines.

We like the 6¼" deep frame better than the standard one. It holds more honey and enables extracting to be carried out more quickly. Indeed, another reason for using Langstroth equipment is that if you have large extractors and wish to use them to capacity, you must use larger frames than the usual British sizes. Otherwise, it will take up to 30% more time to extract the same quantity of honey, and I can tell you after years of extracting, that it is not a job to linger over if you can help it. The frames are put together in the honey farm during the winter months – I can wire 130 brood frames in an 8 hour day!

Choosing an area for bee farming is not easy. The grass always looks greener elsewhere, but I do think that semi-commercial beekeeping can be carried out in most areas of Britain, always assuming that one is willing to move to pollination.

Most beekeepers have little choice of area, having to build up their enterprise near home. But if one is to be a full-time bee farmer then it is important to setup in a really good district not already over-populated with bees. This is not easy.

I think however that the man with a good job and a few hundred hives as a side-line has perhaps the best prospects today. He has a fairly secure income and can easily produce a few tons of honey which will greatly augment his returns. If, however, he goes full-time he will soon find that is takes a tremendous amount of capital just to keep the wheels turning, and even if he does succeed he will be very little better off in the end, when the tax man has had his share.

Labour costs are extremely high today and rising all the time, so that on a large bee farm you must factor this in for each bee hive.

If you add the cost of transport and sugar and the hundred and one other costs that crop up, you will soon find that a very large part of the proceeds go straight back into the business, and that there is not so much left as one might imagine.

The part-time beekeeper does not have such worries, or at any rate they are not so big! In a large business it becomes increasingly necessary to raise the output per man, even if the output per hive has to fall.

I think today it is vital for the larger bee farmer to move his bees to pollination, in spite of any long distances involved. The expenses involved in running the business are so high that the pollination cheque must make the difference between success and disaster some years.

We ourselves, in the spring of this year, plan to move at least a few hundred hives to Kent, as our own raspberry growers are on the whole unwilling to pay a pollination fee.

If it is an economic proposition to move so far, we will move more in future years. We have considerable experience of moving bees in the north, so it should not be too difficult to move to Kent.

Salev forklift (electric).

As labour is so expensive on the bee farm today, there is a distinct move towards mechanisation of all the work involved in moving bees and handling honey, which is a very heavy product.

Forklift trucks are becoming essential around the honey house. They are needed to handle the empty jars coming in and the full ones going out.

They are also needed to handle the big 45 gallon drums which weight 300 kilos when full. Unfortunately, these trucks are very expensive. Electric ones are best for use inside the honey farm buildings. Even small ones like the Sherpa are a great boon.

In the field, the cross-country four-wheel drive forklift is becoming popular too, especially in France and in America where, on the largest bee farms it is replacing the boom loader. The Bobcat, though widely used, has to be towed behind you when you move to the next site, so is not worthwhile unless large numbers of hives are being moved at one time. For smaller numbers the boom loader is perhaps a better aid for the beekeeper but I am told that in rainy conditions they are apt to give a lot of electrical troubles. Nevertheless, I would like to try one, but our sites are not all suitable for its use.

We find in Scotland that for average journeys of 30 miles or so to the moors, the best all round method of moving bees is to keep either 20 or 40 hives at each apiary and to use pick up trucks that will carry 40. With three trucks like this, and a man and a boy on each truck, it is possible to move our 2000 colonies to the heather in three weeks.

Of course, the moves are all made in the early morning, or late at night, in addition to the routine daily work associated with swarm control and removal of flower honey, so there is not much time for sleep for anyone during July!

We have tried a lot of different vehicles on the farm. The Volkswagen 1-ton pick up proved to be the best all round vehicle for many years, but we find it a bit small now. At the present time we have one Mercedes, one 35cwt Volkswagen with twin rear wheels and one Aro four-wheel drive truck from Romania. We fit winches on the front when conditions are likely to be wet.

Extracting

Where large quantities of honey have to be extracted it is necessary to have a well-planned room for the purpose and if it can be kept for that purpose alone, so much the better. If heather honey is being produced this is even more important because of the extra complications in production. Our plant is designed to handle two tons of heather honey per day, or 3-4 tons of flower. We employ only two students for the flower honey, but when we are handling the heather it requires at least one skilled operator and three assistants. We work long hours to

The extracting room with Arbroath loosener.

get the work done as quickly as possible. In the evenings we employ additional staff. The supers are stored in a warm room until ready to be extracted. They are brought in on pallets and the combs are fed into a Penrose uncapper. This is a very fine New Zealand-made machine which uncaps 15 frames per minute. The frames pass through steam-heated blades and are fed out onto a conveyor.

After passing through this tank which removes all the wax particles, the honey is pumped directly to the 300-gallon storage tanks above.

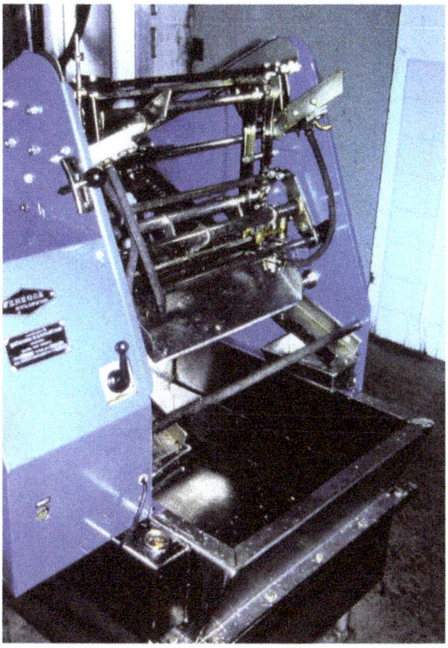

Penrose uncapper. The cappings fall into a Brand type of melter, which separates the honey from the wax. The wax runs directly into moulds.

Brand melter. Flower honey is extracted in a 72-frame Kelley extractor fitted with automatic speed controls, and from the sump is pumped to a baffle tank.

Automatic drive for Kelley extractor.

Kelley 72-frame radial extractor.

Float operated warning bell.

The pumps are all automatically controlled by float switches which come into operation when the honey reaches a certain depth.

Heather honey combs have to be passed through a loosener before extracting and we have found the Sjole model from Norway to be the simplest and most efficient in use. There are other models of French and Norwegian manufacture.

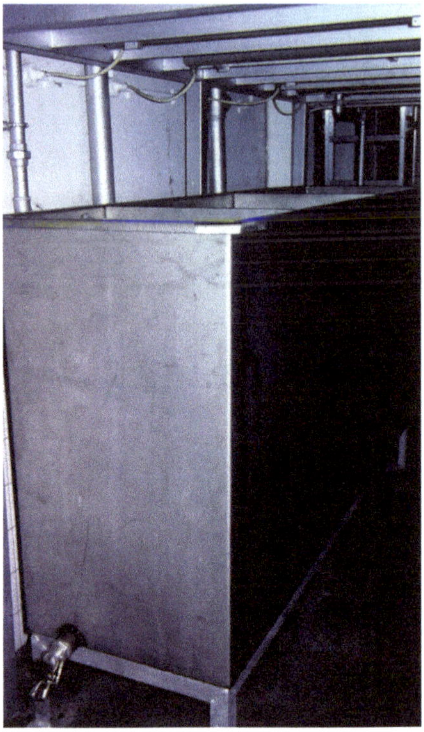

Baffle tank.

Sjole honey loosener.

After loosening, the combs can be extracted in a tangential or semi-radial machine, but not in a plain radial. We have tried quite a few different types of machine, and have settled on the "Versomatic" supplied by Thomas of Orleans, France. We use two of these machines

Thomas Versomatic.

Extractor outlets.

together. They each hold 24 frames. They extract very quickly, they are efficient and easy to use.

As heather cappings cannot be put through a Brand melter because of caramelisation and other problems, they are allowed to fall into the tank below the uncapper where they mix with all the honey from the extractors. The resulting "porridge" is then pumped by a special

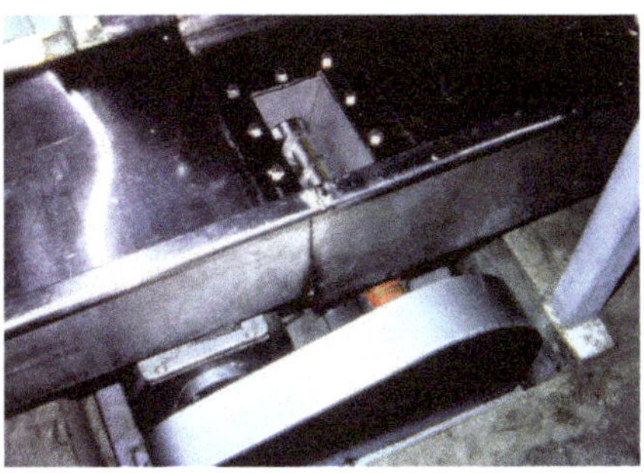

Mono wide-throat Auger pump.

Heat exchanger gear box.

Heat exchanger.

Avamore centrifuge.

Spin float separator.

wide-throat Auger pump through a heat exchanger to a spin-float separator. The mixture enters the spin-float at 120 degrees Fahrenheit.

The honey flows from the machine into a sump, and the dry wax particles fall into baskets below. This American-made machine has been of great benefit to us in the handling of heather honey.

From the sump the honey is pumped to the storage tanks above. Later on it is run into 25-gallon drums for storage. We much prefer the 25-gallon size for

Avery 500kg scales.

Kelley honey sump.

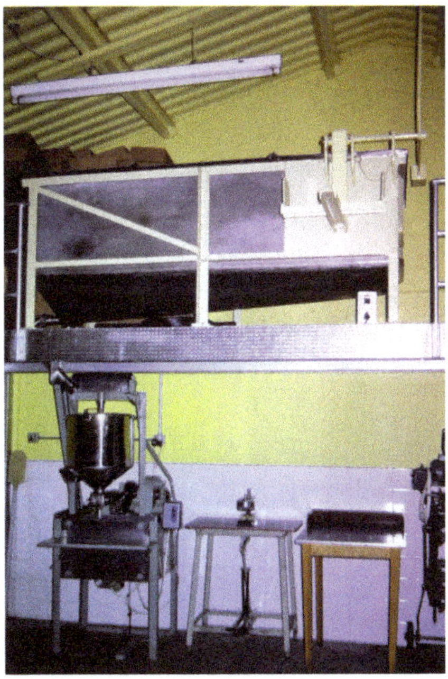

Stainless steel bottling tank with tipper and Centrac Volumetric filling machine.

Labelling machine.

Jars of honey on pallets in cold store.

our own use, but we use the standard 45-gallon size for our export markets on the continent. Our bottling plant is a simple one, but we can pack 2 tons a day in 1lb or ½lb jars and this is adequate for our needs. The Centrac filler is ideal for our purposes and gives exact weight which is very important. A Purdy machine is used to label the jars before despatch.

Bees take a lot of feeding at certain times of the year, and as it has to be done very quickly in autumn especially, we use two 500-gallon stainless steel mixing tanks to make the syrup. The tanks are used with cold water only. 30cwts of sugar is poured into each tank from the floor above and after 2 hours of stirring the sugar is completely dissolved. The syrup is transported to the apiary in tanks holding about 80 gallons. Each hive has a Miller-type feeder which holds 2 gallons. This is the best all-round type for our conditions, and no syrup is wasted, as when lever-lid tins are used.

Sugar hopper.

Two 500 gallon syrup tanks.

Beekeeping, like all branches of agriculture, has to be learned by experience. We are all learning as we travel along, and I hope we will always continue to do so.

End of lecture

The BFA Spring Meeting in Pitlochry organised a trip each afternoon to visit the honey farm in Bridge of Cally.

Bees on pear blossom in Kent – May.

Balintore Castle.

Heather Hills hives at Selling, Kent being moved from unloading central point into apple orchards.

It was decided to experiment with moving the bees from Scotland to the orchards of Kent to benefit from the pollination fees of the fruit ochards there. One hazard that was unforeseen was that the drivers found themselves carrying a load of bee hives along The Mall in the midst of the Queens Silver Jubilee celebrations.

Hives in cherry orchard, Belmont, Kent.

Tummel Bridge heather site.

Apimondia

This year we set off on our world tour in October. Barbara and I left Heathrow for the USA accompanied by Joan and Les Thorne of Wragby (the largest manufacturers of bee equipment in Great Britain). After stops in Chicago, Springfield and Burlington we arrived at Keokuk, Iowa and visited the Dadant Wax factory in Hamilton, Illinois on the Mississippi where we were entertained by Chuck and Arlene Dadant. The next day we flew to Dadant Woodenware factory in Polson, Montana, to see the latest products.

Les Thorne and Chuck Dadant.

Beeswax.

The wax factory in Hamilton, Illinois

Les in in the wax factory.

Jack Allen, Manager, Dadants, Polson.

The Dadant woodenware factory in Polson, Montana (and above right).

We spent time visiting some apiaries in the area and discussed the production of honey with local beekeepers at the factory.

The following day we moved on to Los Angeles where we were met by Pat and Jo Eakle of Dadants' Fresno branch.

We also visited the Sioux Bee Honey bottling plant, probably the largest in the world.

A small Montana apiary.

Joan and Les Thorne, me and Barbara with Pat Eakle, Dadants.

Fred Kruger, Sioux Bee.

Vats of honey.

Honey tanks.

Bottling line.

In the afternoon we drove north by car to see Bill Huston, at Corona, the largest bee farm in California, with 16,000 hives. His crop is mainly orange blossom/sage honey. We saw all the latest equipment for moving beehives to pollination on oranges, apples, etc and photographed the machinery at the factory for my future reference.

Bill Huston with two 180-frame extractors.

Boom loader.

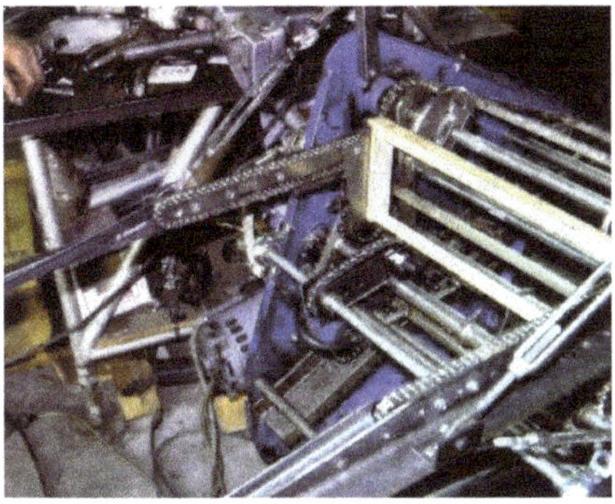

Modified Penrose uncapper.

Knotsberry Farm, LA.

On 11th October we flew from Los Angeles (with stopovers in Honolulu and Pago Pago in American Samoa) to Sydney and on to Adelaide to attend the Apimondia Congress at the Festival Hall which was attended by nearly 2,000 beekeepers. We joined nearly all the lectures, simply as listeners. We met Japanese, German and Dutch honey buyers which later led to sales to Holland.

Adelaide Festival Hall.

Cowan uncapper and extractor.

Home made extractor!

Ecroyd tipper.

Sander.

Pender equipment at the show.

Douglas boom loader.

Pender super lifter.

Olley boom loader.

Penrose uncapper and tank.

We spent six fascinating days here with visits round the local countryside and a Koala park.

From Adelaide we then moved on to Maitland, New South Wales to visit the Pender Bros. factory (where we ordered a large quantity of extracting machinery).

Mr Pender.

Barbara.

Alan Bennet, Pender Bros. with 1-frame 'slinger'!

Boom loader at the college.

College hives.

We paid a visit to the Queensland Agricultural College after flying up to Brisbane the next day and also drove into the country and met Norman Rice, the largest queen breeder in the world, at Beaudesert.

Hives at Queensland Agricultural College.

Norman Rice MBE, queen breeder. *For more about Norm see his book in the bibliography.*

While in the area, we also drove into the town of Tamworth to visit commercial beekeepers, Bob Guilliford, Mr McNally and Mr Keppie. I took photographs of all of these plants.

Bob Guilliford, Les Thorne and Mr McNally.

Mr McNally's Honey Farm, Tamworth.

Mr Keppie's extracting plant.

Mr Keppie's radio-linked vehicles.

Olley's Honey House, Brisbane.

Guilfoyle wax plant.

Guilfoyle's showroom.

We spent two days at the factory of John Guilfoyle at Darra, Brisbane. The next flight was on to Christchurch, New Zealand at the invitation of Kevin Ecroyd, director of Alliance Bee Co. We toured his wax and hive factory, discussed his uncapping machine and problems with it (we use one here).

Kevin Ecroyd.

David Penrose's boom loader

We met local beekeeper, David Penrose to discuss his one-man operated beehive loader and also met commercial beekeepers Harry Cloake and sons.

From Auckland we flew to Fiji where we spent two nights resting! Then we moved on to San Francisco for some sight-seeing. Barbara had many memories from when she had spent time there as a young girl, having been evacuated to the USA from Dairen in Manchuria during WW2.

All in all, a wonderful trip!

Back home, spring losses were again rather heavy, but not so bad as in the previous year. Raspberries gave very good crops in the Mains of Grey and Mylnefield areas of Dundee but not so good elsewhere. Quite good in Kirriemuir.

The Cordingley hive – designed by Francis

Bell Heather

Norwegian visitors, Jordanstone former POW camp

Bell heather looked very good on the hills, but July was very dry indeed, and hot, and it did not yield at all. A little bell-like honey came in at the beginning of August and the Ling started very well. Unfortunately, the rains came about the 12th and this finished the flow completely. The weather did not improve till the heather was past. With the dry weather in July bees had to be fed at that time. It would have been best to have fed all of them, but only about half received a gallon. The rest got none.

Ian Kirkwood and John Holms loading nucs – year old queens in shallow supers

1978

The winter was long and rather severe, so that losses were high once again. In addition 180 live colonies were sold, and this accentuated the losses. 450 hives were also sent to Kent for pollination (all on one 40 ft trailer). 400 were sent to the Bee Farmers Association, and 50 to Mounts of Canterbury.

Losses in Kent were high due to starvation and dwindling in the very poor Spring. Access to many hives was impossible and so feeding could not be carried out properly.

After two seasons in the orchards in Kent it was accepted that with the overheads of accommodation, hire of vehicles and the staff preferring to be working in Scotland, it was an experiment that was not worth pursuing any further.

The raspberry flow was the best for many years, and had colonies been strong a good crop would have been obtained but many colonies were very weak, and strong ones had to be split. June was good and all empty hives were filled by the usual date. July was poor, but there was a good flow from thistle or knapweed in the Errochty glen, but Bell yielded poorly or not all, although there was plenty blossom.

Hundreds of hives at farm ready to move for pollination in Kent

Loading 450 hives for Kent in April

40 Langstroths, Deeside

The Ling came early and blossom was very good, but heavy thundery showers every day or two spoilt the heather, and it was all over by the 28th August.

We had a group of visitors from the north of England who came to see the farm in August.

Bee boles spotted while on holiday in the Lake District

1979

Norman with Wilson Carne

In June we had a visit from beekeepers from the Elgin area and in July a large delegation from Belgium. Norman and Heather Rice came to see us while travelling on to Canada and America.

This year the heather crop would be augmented by syrup fed in August. The winter was long and severe, and what little Spring there was came very late indeed.

The result was a very heavy loss of colonies in April, and a dwindling of numbers in practically all the hives.

200 colonies were sold to Mounts before the full loss was realised. To counteract this we bought from Malcolm Walton, Southampton 275 good colonies of Italian bees. These saved the situation. Being very strong they were used to boost the weaker hives, and to make a great many nuclei and artificial swarms. Some still stored over 100 lbs of raspberry honey – even after splitting. The raspberry flow was the best for years but was exactly three weeks late.

The Wallaces, Quince Honey Farm, South Molton, Devon

The Bell did not yield, and it was too cool for clover. Weather was very unsettled all summer and there was heavy rain almost every second day in August.

1980

This year the heather crop would be augmented by syrup fed in July/August.

We had two large groups of beekeeping visitors in August, firstly of Norwegians and then from France.

The winter was not bad, but losses were again heavy. This was due chiefly to bad mating the previous poor summer, and also to the poor work done on the hives by some of the staff at that time.

With rising costs, and 21% inflation, it was decided to cut back the number of hives. and the best 370 were sold, so that the remainder were on the weak side. However, a wonderful spell in May helped them to recover. After that there was no more decent weather, and the whole summer remained cool with a great deal of rain almost every day at some time or other.

The Raspberry flow was quite good in Blairgowrie, but hives had all been split severely and a crop could not really be expected. In fact, about the middle of June we felt that no crop could be expected during the whole year, hives had been so severely split.

However, there was a good flow from something in July, especially at Ballintuim and Ling yielded quite well where hives were strong enough to store a surplus.

Glenisla

1981

Apimondia Tour – Acapulco with Joan and Les Thorne

Agricultural College, Ohio – Les Thorne, Lawrence Goltz and Dr Tew

The A I Root Co

Walter T Kelley Co.

83 year old Walter Kelley at his desk

Kelleys making queen cups

Kelley cappings reducer tank

Kelleys making queen cups

Kelley mini-uncapper

Acapulco

Apimondia

Maxant spin dryer

Fields of Ambrosia

Cartwright Plastics

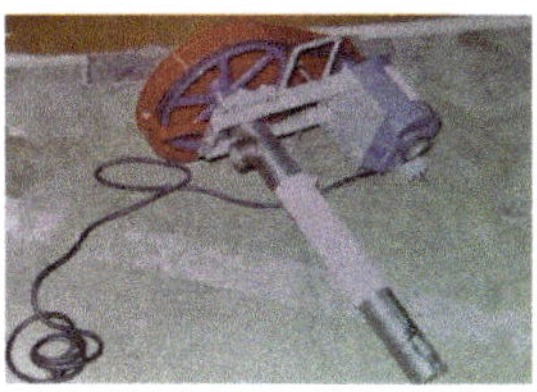

B16 Maxant cappings pump

Barbara with Nora and Cecil Tonsley and Les Thorne

Korea trade stand

Mexican trade stand

Rietsche trade stand

Bees near Acapulco

Acapulco Honey farm examining nucs

Acapulco Honey Farm – 3 extractors

Acapulco Honey farm hives for repair!

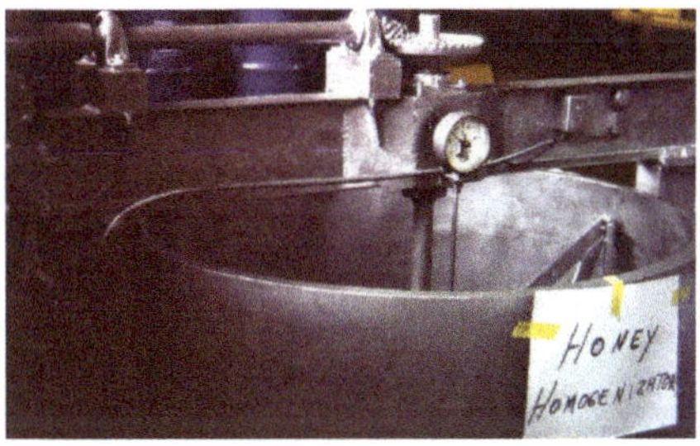

Very large homogenizer

I felt I had achieved as much as I could in building up Heather Hills Honey Farm and began to consider early retirement. The owners of a local spring water company were interested in the business and the buildings and 90% of the remaining hives were sold in May to their subsidiary company, Honey Farmers Ltd for a very good price. 154 hives were retained for my own use.

1981 was a season which started badly and finished with a glorious August. Spray damage reduced the raspberry crop considerably, and there was a lot of Spring dwindling, perhaps due to pollen shortage.

July was fairly good with big crops of bell-type honey in some apiaries. A lot of this would be thistle and knapweed and clover. Other apiaries did badly. Strathardle was all excellent, but Ceagnlinne rather poor. Bridgend was very good, Balnald excellent.

The Ling lasted the whole month and yielded steadily giving very good crops – the best colonies gave around 120 lbs ling alone, and over 200 lbs, for the season. It rained only about two days in the month, and it was very hot. The hills became very dry but the flow continued because of the heavy dews. The ling flow was perhaps the best I have ever seen. No sugar was fed this summer and very little in September. All were very heavy.

The new owners of the honey farm also did well, with the biggest crops in Aberdeenshire.

Myself in sporting gear with the honey farm in the background

CHAPTER 4

Retirement

1982

Embarking on my first year of retirement, unfortunately the losses were fairly heavy again after the coldest December in living memory.

The early days of Honey Farmers Ltd – Iain Mackenzie, Allan Small and Quentin Henderson, used by kind permission of D C Thomson.

After January the weather was not so bad, and the raspberries were a full two weeks earlier than usual. The flow was finished soon after the middle of June. Rape increased in acreage, but did not seem to increase the crop, though the pollen was certainly very useful.

July was a very good month in Scotland, though poor in England, and very high temperatures were recorded, 80°F. and upwards. The drought became severe, but clover yielded well, especially in West Scotland where some very big crops were obtained. The bell did not yield much in Perthshire, but gave big crops in Angus, where Bert Mason had one of his biggest crops ever. In Perthshire there was a flow from other flowers instead, but no bees were seen on the bell heather.

Raspberries at Blackhills.

Errochty Loch.

Ling began to yield well about July 24th, again much earlier than usual, but about 4th August heavy rain came on and the rest of August was too cold and wet, so little more was obtained.

With August being so bad for matings, fairly heavy losses are anticipated in Spring of 1983.

Ling – Moulin Moor.

1983

The winter of 1982/83 was, for a short period in December, the coldest again in living memory. However, after January it was fairly mild, and easier on the bees than normal. Losses were very reasonable, and colonies came through strong in most but not all apiaries.

Spring was the wettest in living memory, and the rains continued right through till June! Fortunately, the rasps were two weeks late, and there was a good honey flow. Several of the best colonies stored over 100 lbs. surplus. The bell flow was poor because of drought conditions which continued on into August. The weather turned in August to the hottest for years, until it broke about the 20th. With the flow being late from ling too, this finished the season abruptly. A few nights of rain earlier would have made a tremendous difference to the crop.

Most colonies needed little feeding, and only division board feeders were used, twice.

All honey was extracted, by arrangement, at the Honey Farm. Ling was done in two days only.

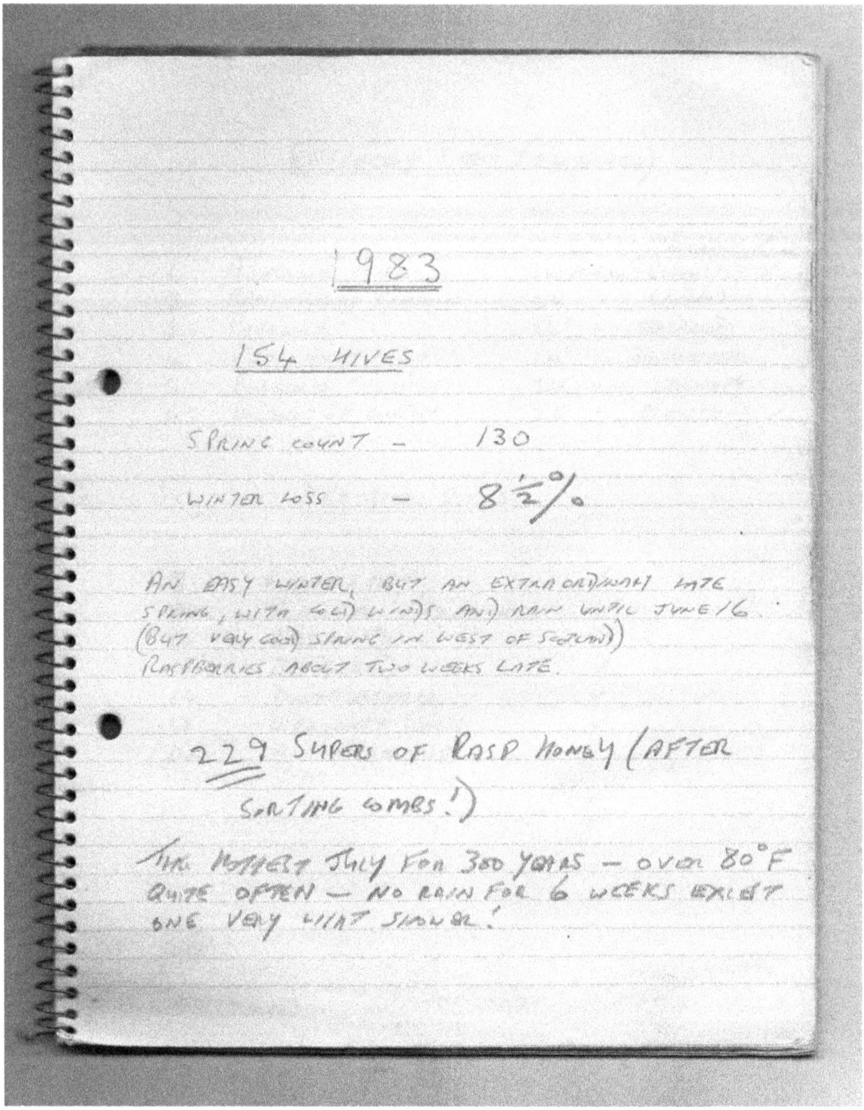

Apiary Record Book.

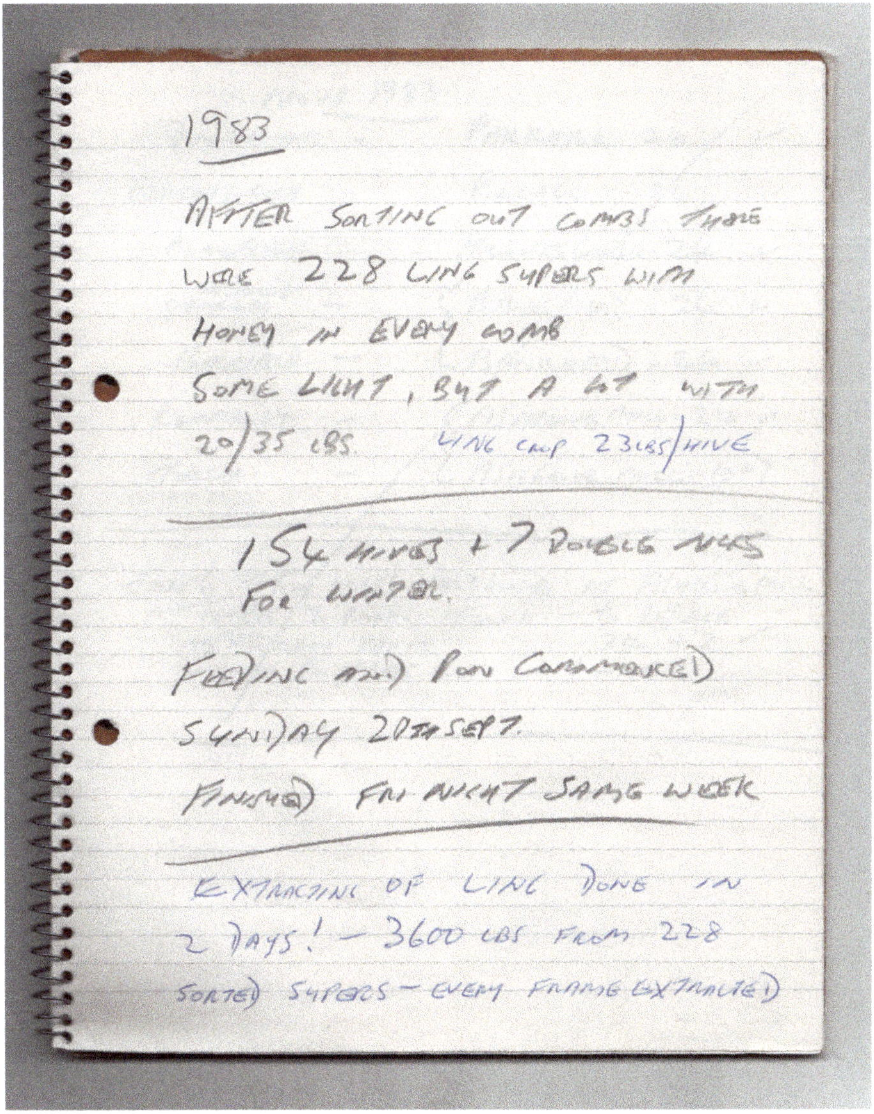

Apiary Record Book.

1984

This was an extraordinary season. Certainly the best summer in living memory, and unlikely ever to be repeated!

The sun shone almost every day from Easter until late August, and rain fell on only four or five days during this very long spell.

In spite of the severe drought which dried up many of the hill streams and reduced the level of the rivers to unprecedented levels the honey flow did not suffer in the Glens although it got too dry in lower areas. It was possible to walk across the river Ardle in many places without getting one's feet wet! The biggest yields were obtained at Ballintuim (as in 1959) and Blairfettie was also very good.

Oilseed rape came in quantity into the Blairgowrie area for the first time (although it had been grown at Rosemount for a few years previously).

The winter was very good for my bees, and losses were minimal. All hives were restocked by the nuclei (overwintered in two shallow supers). Most of the colonies were good and strong by May, much

Oil seed rape.

Balnald Apiary.

earlier than usual, and 5 of the 6 apiaries were within reach of the rape, although some were a good half-mile from it.

The rape flow was very good. So was the raspberry. July clover and other flowers yielded well, although bell was as usual not so good. August was excellent, but the ling honey was thin and extracted easily after loosening. In fact, it was extracted in a radial machine. The ling was extracted at Heather Hills Farm by Fiona, and one other helper.

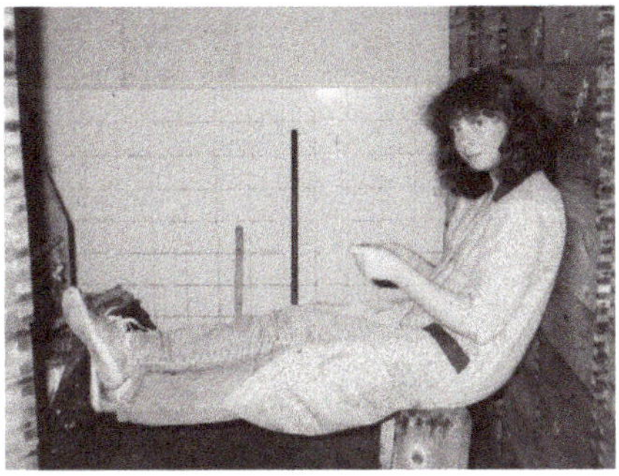

A short break.

Ling supers on truck.

The 5 tonnes were extracted in four working days (8 hours for helpers and 14 for myself).

The BFA Conference was in Thetford in May and I went with Bob Cousten and Norman. This year I took on the Chairmanship of the Bee Farmers' Association.

Norman Stewart at Runavey Apiary.

1985

1985 was a terrible season, although I did quite well.

The winter had been a very mild one, one of the easiest on record, which coincided with the building of our new house in Kirkmichael together with a new honey shed. Although my dreams of returning to the village were coming true I had never thought I would live on the banks of the Balnald Burn I had fished as a child, listening to the rushing waters at the foot of the garden. Work went on all winter.

Bees came through strong, but the weather never warmed up at all for long. There were a few nice days in April and May, but the cold winds persisted. The flow from rape was good, and also reasonable from the raspberries, but from the end of June onwards things went from bad to worse, and July and August were both miserable with hardly a day free of rain. The bees were all taken up to the hills as usual, but it really was a waste of time. They did almost nothing and had to be fed often just to keep them alive. Mating at the heather was out of the question. The weather did improve a little at the end of September, but it was too late, even though the heather still looked reasonable.

Fortunately, with my present system of keeping year-old queens in nuclei (shallow supers) all hives had a queen before autumn. Bees looked quite good going into winter, but there was a great shortage of pollen until the end of September.

Building The Sheiling

...and a new honey house.

The Sheiling, Balnald, Kirkmichael'

Group including some individuals who formed The Council Of National Beekeeping Associations in the United Kingdom (CONBA) taken at Perth, Scotland during the Spring Meeting of the Bee Farmers' Association, on 19th March, 1985. Left, back — R. Rawson-Davies (Secretary BFA); Alistair Lilburn (President SBA); Robert Couston (SBA; CONBA); George Smith (SBA; CONBA Secretary). Seated front — J. McIlroy (Ulster BKA; CONBA); Athole Kirkwood (Chairman BFA); Commander Eric Verge (Welsh BKA; CONBA Chairman); Michael Coward (Secretary BBKA; CONBA).

Photo: Cowper & Co., Perth.

27

BFA Spring Meeting at Perth 1985 (Photo Cowper & Co).

What Other People Think

Letter to *The Scottish Beekeeper*, January 1986

Dear Sir,

I do not often put pen to paper but after the past dismal season a few comments might be of interest. Bees wintered well here, after an easy winter. For a second year in a row colonies came through very strongly and losses were few.

1984 had been a wonderful summer in which it was difficult to do anything wrong. Just as in 1959 anyone could have got a crop, and my own yield was no less than 25,572 lbs from 154 colonies! Just an average crop by American standards, but a lot better than I have ever done or expect to do again. 1985 started even better!

By the end of June, with the same number of hives I produced 9,400lbs of raspberry and oil-seed rape honey. A few colonies produced over 200 lbs each. I began to wonder if the great flow might continue but of course it didn't. Raspberries yield well in quite moist conditions but July flowers do not, and it soon became necessary to feed.

There were some very brief dry spells in July and August and a little honey was produced. 470 lbs in July and about 600 lbs in August.

Running so few colonies nowadays (I used to run over 2,000), I was able to keep every colony in peak condition. Normally this ensures a heather crop even in quite poor seasons, but not this time! In 40 years of beekeeping I have never seen such a season, and do not want to see another like it. I finished up with an average crop, but it was a pretty miserable job producing it!

Yours faithfully,

A.M.C. Kirkwood, N.D.A., N.D.D.

1986

This was another very poor season. Losses of colonies were very heavy after the bad summer, long winter and very late spring. Only Ashgrove Mill was strong.

All others with big losses. Had the bees been in good condition there would have been a good raspberry crop. Rape did not yield well however, probably due to the cooler weather. After June it was too cold nearly all summer, though there was one period in June when it was about 80 degrees F. for a whole week. After that the wind got colder and colder, and only on a few days here and there were bees able to fly at all. It was not all that wet, but with the low temperature many queens did not mate for a long time, and some not at all.

The flow in August was very poor, except at Blairfettie where there was some honey in August – possibly honeydew? There was no honey at all in Glenshee or Glenisla.

Fortunately, the weather did improve in late September and bees were mostly able to fill the broodchambers. More importantly they got a lot of green pollen. There had been a great shortage of pollen before this.

I called this the "year of the cold winds". Good for hay. No use for bees.

The BFA Conference at Great Malvern was the last I attended with Norman who was not keeping good health.

Ashgrove Mill.

The Hive of the Future?

A.M.C. Kirkwood

Excerpt from an article in *The Beekeepers Annual 1986*

Before the discovery of the bee-space bees were kept in a variety of hives, some very intricate, and others very pleasing to the eye. With the introduction of removable frames there was a lot of scope for development and no shortage of designers. The British Standard frame came into being in due course, but for a long time there was no standard hive, and the result was a complete non-interchangeability. Later on there came the WBC hive, but many copies were made by carpenters who did not realise that exact dimensions had to be adhered to. In most cases beekeepers ended up with several different patterns of hive, and this fact certainly mitigated against expansion of their apiaries.

In due course the National hive appeared, standardised by the Department of Agriculture, and this was perhaps a step in the right direction, but it is interesting to note that nobody ever built up a really large bee farm with this hive, so far as I am aware. Of course it was not designed for really large scale work. It was a very cheap hive designed for the British Standard frame and it made the transfer of bees from the older hives easy to carry out. It was also a handy hive for moving about, compared with those that went before. For a beekeeper with a few hundred hives it was, and still is, quite a useful hive. The Modified National is in some way a nice hive too, but suffers from the same disadvantages.

The hive of the future should have larger frames, both in the broodchamber and in the super. Larger frames mean less expense in building a hive of certain capacity. The smaller number of frames makes for faster examinations and a very much quicker extracting.

The hive of the future should of course be of the single-walled type for the sake of economy, not double-walled at the ends to accommodate long top-bars. It should have a top bee space which makes for much easier handling, and hand holes on all four sides.

The roof should be deep enough to cover these hand-holes, but not so deep that it is unwieldy. It is best covered in galvanised iron which stands up to wear better than aluminium or zinc.

Looking at these desirable features one can see that the hive of the future is here today, and has indeed been around for a very long time. I refer of course to the Langstroth, which if it had been introduced instead of the National would by now have been standard in this country. The National hive has unfortunately meant the perpetuation of the old BS frame which would have disappeared long ago. The Langstroth is just about the only hive used today in the major honey-producing countries.

I became very friendly with the late Wm. W. Smith of Innerleithen, Peebles-shire, a marvellous man, and a great beekeeper. He had brought out the Smith hive, and got huge crops with it. But I had also entered into a long correspondence with R.O.B. Manley and A.W. Gale of Marlborough. These two had very different ideas on the subject of hives. R.O.B. Manley's books were an inspiration to me, but A. W. Gale used B.S. frames in a 13-frame hive, and I felt at the time that it might be best to spend some time with him. I went to Wiltshire in 1950, and in just one summer learned a great deal about large-scale bee-farming.

During that next winter, after I returned from Marlborough, I spent all my spare time nailing up Smith hives. Only 50 at that time, and it seemed a lot. Little did I know that I would be nailing up a further 3000 in the next 20 years!

All my equipment was bought from Lees of Uxbridge or E..H. Thorne of Wragby. Both gave excellent service, and their equipment was a very good investment. The first Smith hives were duly stocked with French Package bees, obtained at that time from R. Steele & Brodie. At that time I lived in the West of Scotland, and my apiary was at Bridge of Weir in Renfrewshire. Even though it was a wet district the French bees did well, and nearly always stored a crop.

Each autumn I ordered up more Smith equipment, and each time I thought "these supers are too small", but when on a treadmill it is difficult to stop and view the position objectively. By the time I had 1100 Smith hives it finally hit me. I had to get a better hive for the job I was trying to do!

I do not like double broodchamber systems of management, though I used the Demaree method of swarm control with success for many years. I do like to keep the queen in only one box at all times.

I experimented with the Langstroth Jumbo hive for a year or two, but found the big Modified Dadant brood frames too deep for heather honey production, even when I reduced the number of frames with dummies. It was too deep. I tried the Langstroth, using 9 frames and a division board in the broodchamber, and very quickly found that this was ideal for my purpose. The decision to move over onto all Langstroth equipment was quickly made. The 1100 colonies in Smith hives were sold or transferred into the new hives. The complete change took about five years, with an expansion programme thrown in.

By 1976 there were 2050 stocked Langstroth hives, and the benefits immediately became clear.

First of all, I was able to purchase from overseas, particularly USA, Australia and New Zealand, plant and machinery designed to handle the bigger frames of honey. Previously I had imported similar machines and had spent a lot of money altering them for British frames. After alteration the result was that I had reduced the output of these machines by some 30-40%. Not exactly sensible!

In the field, we used 9 frames in the broodchamber instead of the normal 10. We found this big enough for our dark bees, and instead of a division board we soon found a very good plastic 'dummy' feeder. This is left in the hive at all times and holds about ¾ gallon syrup. These are obtainable only in Langstroth size. They are made in USA. Bees will take syrup from them in quite cold weather when they will not come up to top feeders.

In the supers I have always used the Manley close-ended extracting frame, which I like very much. Because we are using them in a super of Langstroth width and not in a Modified Dadant, we have found it advisable to make the sidebars one sixteenth inches wider than standard. This slightly wider spacing makes the super ideal for eight frames. It is interesting to note that the Australians have also done this for practical reasons.

The benefits of these large extracting frames are less handling and much faster extraction. Also, a lot less nailing up. We like the 6¼" depth frame best. The supers hold about 40lbs when full, and weigh about 56lbs gross. It is of course possible to pass these frames through an automatic uncapping machine at 15 or more a minute. This is not possible with British Standard equipment.

With a crop of about 50 tons, we could do the extraction in about half the time it would take with the smaller British frame. It also takes half the labour and half the power. Extracting large quantities of honey is a very hot and tedious job. I know that when I was employing a staff of 10, not one would volunteer for the inside work if there was a chance to get out into the field with the bees!

Ed. The bee-daughter can testify to this! From an early age and as soon as she was able to stand on a box at Achalader loading the frames into the extractor, she had helped during the summer weeks and endured the hot sticky conditions in the extracting room. The men would be out in the fresh air, admittedly tackling the heavy lifting of supers, while the women were invariably in the factory sweltering over the machinery, unable to open the windows as the bees outside, smelling the honey, were ready and eagerly waiting to descend upon us!

Nowadays I only keep 154 colonies of bees, but the advantages of the larger frames are still very obvious.

In 1984, with the help of Fiona, we extracted over 25,500lbs of honey in a matter of days. In earlier years it would have been months! Of course, this was a most exceptional season, and I doubt if it will ever be as good again. It is not often that one averages over 165lbs on every hive in this part of the world. Although I have several times had over 100lbs.

However, the really great advantage of the Langstroth hive is that any beekeeper using this size will in future years be able to make full use of any new labour-saving devices that appear on the beekeeping scene in any part of the world, and there must be lots of good things still to be discovered.

Why not change to the hive of the future, today?

The bee-daughters's wedding day.

Glen Derby.

1987

This was another disastrous year for most beekeepers in the UK, but Perthshire, and Inverness-shire (Hamish Robertson of Struan Apiaries) did quite well. There was a good crop of heather in the New Forest, and in parts of Derbyshire, but the rest was mostly very poor.

Bees were not very strong in Spring, and the rape did not yield well anyway but the raspberries did well as usual, and increase was easily made. The crop was good.

July was a bit too cold and wet for good crops, but there was a good crop at Blairfettie – possibly honeydew from the pine forest. Other sites were not so good. August was not a good month for weather. Blairfettie gave practically nothing, perhaps too far from the ling. The other sites did moderately well. Glen Derby bees did well, right in the middle of the heather, though bell was disappointing.

Ed. Although it was obviously a mixed season, Dad gave a very positive talk to the Glasgow SBA in September of this year which gives an interesting indication of where he felt himself to be at this stage in his career.

Scottish Beekeepers Association talk 1987

A.M.C. Kirkwood

I spent about twenty years of my life on the south side of the city, went to school and college here, and started beekeeping here too.

I first took an interest in bees in 1944 – seems a long time ago, and I begin to feel like a historian! Having kept bees for 42 years and made my living from beekeeping for 33, I suppose I can claim to have been reasonably successful at my craft.

Bees were difficult to obtain during the war but I when I managed to obtain an 8-frame colony from Burgess of Exeter, which arrived on VE Day, I will never forget the thrill of feeling the warmth and hum of these living creatures – it had all been theory up to that day. The bees were safely installed and I was a beekeeper.

District/Climate

As my parents moved around, I did too, along with my bees. I produced some nice lime honey in the city, clover honey in Renfrewshire and lovely ling honey at Loch Thom above Greenock. The late Jack McGregor of the West College used to say that there were pockets in the west, where bees would do very well. In my experience this is perfectly true, but the over-riding factor is the rainfall! In the west one must expect the occasional disaster because of the continual rain. This is not quite so usual in the east, but still does occur at times.

The aspiring beekeeper must get the best possible conditions for his bees. At one time I seriously considered moving to Galloway or to Devon, but decided, in the end, to go back to the Blairgowrie area, where my father had been born. I had been evacuated to my grandmother's during the war and knew that bees did fairly well. I had even been to film shows in Kirkmichael given by the late Joe Roger, Beekeeping Adviser, Perth, (who was in later years joined by my friend, Bob Cousten, the East of Scotland Beekeeping Adviser), but I was still at school and no real interest developed until I was back in Glasgow.

After qualifying as an agricultural adviser and working at this for about three years, I moved to Blairgowrie in 1954.

We cannot all keep bees in the raspberry areas. The hive density is rather high already, but in my opinion it is the best honey area in Great Britain. Oil seed rape has also helped tremendously.

Essentials

The essentials for beekeeping are capital, knowledge, fitness and experience.

A lot of capital is required today. One should certainly have 400 colonies bought and paid for before embarking on a full time job. You also need money to live on, so really it is best to keep a full time job as long as possible. 100 hives and a job is a good compromise, or you get to the stage that I was in in 1945 when I had to sell 2 gross empty jars to pay my rent.

You need some buildings, but these should be as cheap as possible. They produce nothing. All available money should go on the bees. Equipment can be fairly simple at the start – I did about 4 tons in 1955 with a hand-driven machine and a couple of ripeners.

You have to be very fit. Summers are just continual work and it is a very heavy job if you are lucky enough to get big crops.

> Ed.– Dad ended up with a very painful tennis elbow as a result of years of repetitive work including emptying plastic buckets of honey into drums!

R.O.B. Manley used to say that "experience is a great teacher but the cost of tuition is rather high". You do need some years' practice and you must visit and work for other beekeepers.

Bees

My first colony of bees cost £7.10/- including carriage paid by passenger train from Exeter!

Over the years I have seen good and bad bees. I do not like Italian bees, but some do.

Hives

You will need feeders, excluders, frames, stands, clothes, veils, hive tools, etc.

Books

Books can be useful and Manley's books were like a bible to me and I still like to read them again and again. I corresponded with him for years and gained much helpful advice and healthy warnings he was positive I would ignore!

Now, we have moved 34,686 hives to the heather since 1945 and I still feel the same anticipation whenever I arrive at an apiary and can't wait to see what these incredible creatures have been up to while I have been elsewhere.

Ashgrove Apiary.

1988

Another strange season. Bees wintered pretty well, and losses were made up with overwintered nuclei. The 100 colonies taken from Honey Farmers never did as well as my own bees, and it is doubtful if they made even 2 tonnes.

May was a fairly good time for bees and June was superb with very high temperatures almost every day. It was easy to make increase as the flow was very heavy, and almost every queenright stock was split in two, making nearly 500 in all.

July was disastrous with heavy rain and cold winds every day. There were very few periods when queens could mate, but some managed it. Luckily most queens had mated at the end of June.

August started well, with a few very nice days, but after a week it all went wrong and the rains came again and again. What should have been a good crop was washed out. When better weather came in September it was too late for this area

My great friend Norman Stewart of Glen Lyon Apiaries died in Aberfeldy. He had been involved in beekeeping since he was a boy and was the honorary president of the Fortingall Beekeepers Association.

Alrick Apiary.

Norman Stewart 1968.

Norman was born in Perth and educated at Perth Academy before serving as the Cooperative Insurance agent in Aberfeldy from 1933 to 1971, only interrupted by WW2, where he served in Italy and North Africa. He was also on the beaches at the D-Day landings in Normandy in 1944.

Norman and I spent countless hours working together at the bees on the heather moors and attended many conferences, our last trip being to Great Malvern.

1989

Return to Heather Hills Honey Farm

This was a very very good season in most of England and in the West of Scotland, and in parts of NE Scotland, but in Perthshire and Fife it was too dry from mid-June until early August, and then it was too wet. In June and July temperatures here went over 80 degrees F. quite often. Bees built up very well.

On 10th May Honey Farmers Ltd had a Liquidation Sale, and my son, Ian and I bought 30 poor live colonies and about 170 empty hives full of moths. Also, a lot of plant and equipment.

On 23rd May we bought the buildings back, but by this time a lot of the hives had been sold elsewhere. (Only about 160 hives had bees in them at the time of the auction!)

The summer was spent building up numbers and cleaning up equipment. 734 colonies were taken to the moors. Other equipment was purchased to bring the number of hives up to 786.

Three apiaries were established on Mar Estate, by Braemar, however the other sites had already been divided up and I was unfortunately too ill at that time with bronchitis to focus my energies on the auction.

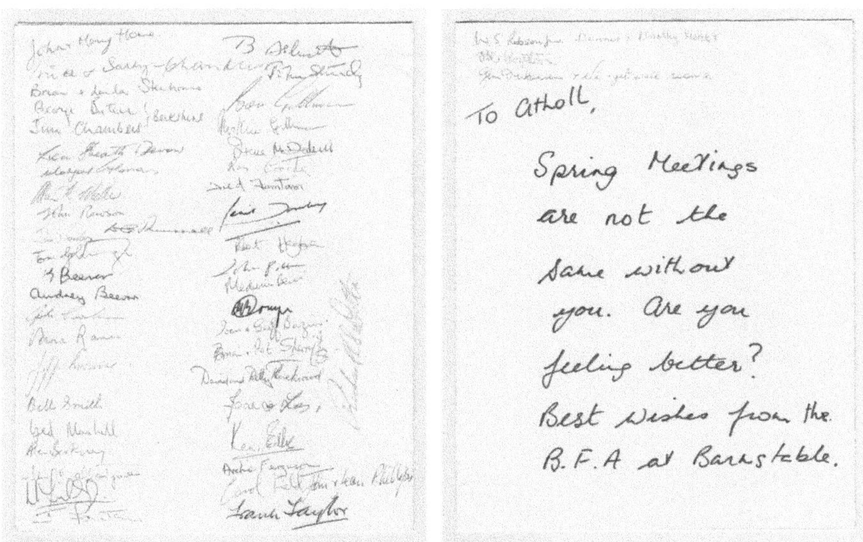

Message from old friends at the BFA Spring Meeting, Barnstaple

1990

My return to Heather Hills Honey Farm was fortunate in that the season was good right from the start. The winter had been easy once again. There was a reasonable flow from rape, and a good flow from raspberry. There was some clover in the hill areas, and a very good flow of bell, although strangely the bell was very light in colour. Not port-wine at all, but it did have the aroma and flavour. Glen Derby was especially good for bell.

Extra equipment was put together to raise the number of hives to 786. During the summer all the main rooms were painted by the student help.

The ling crop was fairly good, but heather at Errochty Loch was diseased, and the crop there was very poor, which is most unusual. Further down the valley there conditions seemed to be fine. Altogether a very good result.

An Achilles tendon injury, sustained while I was trying to push the truck out of the mud at a boggy apiary, meant that full time bee work was really too much for me at this time. Being encumbered with a walking stick, however, while visiting Catriona and her husband in Milan, was very useful in fending off pickpockets when they attempted to steal my wallet. I managed to knock it out of their hands with my cane and send them scarpering!

1991

Kirkmichael Honey Farm

Bernard and Mark Noonan, who lived locally, were seeking a rural business opportunity and entered into partnership with Ian in the spring so I was able to retire for a second time.

I worked full time until end of active season, but did not do any early mornings, except with my own bees. Their crop was around 24 tonnes with an excellent ling crop.

Lorna McGinn, Morag Davidson, Joyce Wilkinson, Sarah Kirkwood and Sally Fenner with Ian Kirkwood and Bernard Noonan standing.

Some apiaries wintered quite badly, but overall the position was fairly good, and losses were soon made up. The summer after June was very good. In June rasps yielded well in spite of the poor weather and lots of rain. Errochty Loch apiary seemed satisfactory this time, after the disastrous time the previous year.

I extracted all my honey at Cally but decided to produce all ling in the form of cut comb and sections. Some crates of sections were 90% full, but only about one crate filled on each hive, even though they were strong.

I bought in September a Suzuki van and small trailer for my own bees.

I attended the BFA Conference in London with Bob Cousten.

Ed. In 1991 Dad was delighted to be made an Honorary Life member of The Scottish Beekeepers' Association, the first to acquire this status, in "appreciation of the time and effort you have given to the furtherance of beekeeping in Scotland and in particular the support you have given to the SBA over these forty years".

1992

Kirkmichael Honey Farm

This was a very easy season as regards work! I helped to clip the queens in May at Heather Hills Honey Farm, and gave a few days work later, helping to move some of their bees.

I extracted the flower honey at Kirkmichael, but the drawn comb ling supers were done at Bridge of Cally, and this is certainly the easiest way to do it.

The flower honey season was better this year, but bell heather honey was not very good, and ling was well down on the previous season.

Varroa disease appeared for the first time in SW England, and later in the season was confirmed in Wales, Kent, Surrey, Lincolnshire and other southern counties.

Francis Cordingley and Fiona.

1993

The season was reasonable, but poor at the beginning of August. I had to take the heather honey off rather early as I was off to China on holiday, but this did not much reduce the crop.

There were 4 supers of heather honey for extraction, and this was done at Cally.

Ed. Our parents travelled extensively at home and abroad visiting China and the Far East, New England, Canada, Spain, the USA, New Zealand and other far-flung destinations, together and with friends, and if ever there was a word of a honey farm in the vicinity dad would be sure to hunt it down!

Chinese honey farm.

On holiday with Kay and Colin Campbell.

Holiday to the Far East.

1994

The winter was very long with hard frost starting very early in November. There were very few days that bees could fly in early Spring. Spring came very late and when it did it rained a lot.

Losses of bees at Heather Hills were around 40-50%. All down the East of Scotland losses were very high and many were almost wiped out. Nevertheless, my own bees did very well. The Summer when it came was good but short.

50 years of beekeeping and still smiling.

1995

Bees kept now for 50 years!

About 20 swarms or nuclei were sold to Heather Hills Honey Farm in June. This was a wonderful summer after a slow start.

The crop was not nearly so good in Aberdeenshire, and there was a poor crop in France, so that demand from Germany was good. Bert Mason sold honey to Germany in bulk. They wanted 20 tonnes but he could not supply all that.

Varroa was confirmed in North Yorkshire for the first time. My good beekeeping friend, Francis Cordingley, who sadly died this year, had kept his bees in that district.

1996

This was a very disappointing year.

The winter was not too bad but went on too long and the Spring was very late in arriving. The whole season was about a fortnight behind – a very rare occurrence.

The bees that were lost in Spring started it, as I underestimated their stores. Again, I lost 4 colonies in June when very little was coming in. This was just due to carelessness!

Bees should have been fed heavily in June and again in July at the bell heather as so little was coming in. It was difficult to see why things were so poor, but it was a cold dry early summer, and there may have been insufficient moisture.

Queens mated well however, so no problem there, and it was easy to make up the losses in early June. There was little or nothing from the bell heather.

At the ling time the bees were able to put in a good bit for themselves, but they were very loathe to build comb – results very poor, and worst for years.

Visited Bert Mason's hives in Glenesk – all had 1–1½ good supers (on drawn combs)!

Bert Mason 1996.

Glenesk.

Ed. Dad of course kept in touch with his many old beekeeping friends Richard and Joyce Wilson from Wales, Brian Sheriff, Willie robson and Hamish Robertson among them, and was always very much up to date with the beekeeping news.

1997

This was a very good year indeed.

May was a good month with a good crop from the early rape, but June was disappointing, cold and damp, so that the raspberries did not get much of a chance to yield.

However, as in some exceptional earlier years, the flow began in earnest about 1st July. As we were away to Italy the second week, I decided to leave the bees on the rape till later than usual, and they did very well indeed during that period. (I was told however that they would have done just as well up the hill!).

The bell appeared to yield for a very long time – right into August in fact or else there was something similar yielding, because when it was time to extract the ling it was very thin, and most honey came out easily in a radial machine. (This has happened before in very good years).

1998

This was a very strange year.

The bees all wintered well and were very strong in Spring. The result was a wonderful crop of rape honey in May, 46 lbs from every hive. June was not good, and each hive gave only another 21 lbs. That is 67 lbs total per hive – a record probably. (Best for May only).

Eddie and Audrey Millington (Eddie always thought I somehow resembled a bee!).

Off to the golf with childhood friend Alf Inshaw.

After June the weather got worse and continued poor right into October. There were two very hot days in mid-September, but too late for the bees.

Had the bees been on drawn comb they might have made about 20 lbs of ling each, but because they were working for cut comb and sections any honey they got, went into the broochambers. Not a lot of feeding was required for winter, but some feeding had to be done in August to keep the bees alive!

Ed. Close childhood pals, Eddie Millington and his wife Audrey, Alf Inshaw and his wife Anne, and Kay and Colin Campbell were lifelong friends with whom Mum and Dad continued to keep in regular contact.

1999

Another strange season.

Spring came late. There were a few amazingly hot days here and there, but mostly cold and wet. May and June were both poor with very little decent weather. Mid-July was quite good, and bell yielded well with very good crops in some areas.

In early August the weather became very hot and dry. So dry that even with the odd day of rain the ling yielded very little. The poorest ling season for a long time. In late August/early September it was close to 80 degrees F. on several occasions even while blowing a gale! Most unusual and no good to the bees.

Weather continued mild right until the end of September, so it was good for feeding.

Beekeepers in West Scotland, and indeed West England did pretty well, as they got rain at the right time. For us it was almost a disaster.

2000

Another poor season for my beekeeping, but son Ian did better than in the previous year.

12 colonies were sold to Heather Hills Honey Farm in spring.

By taking the queens away in early June and making up nuclei, I stopped swarming all right (important as I was away in Canada

The Rockies.

and USA), but the colonies were so weakened that the crop was poor.

The early part of the season was very poor, and what little rape and raspberry honey was gathered, was nearly all eaten later by the bees.

There was no heather honey at all in West Scotland, and it would appear that the Heather Beetle is causing increasing trouble there.

Varroa strips were used for the first time in my hives, Apistan being used in September, October and half November. There was no sign of Varroa but better safe than sorry.

2001

Yet another poor season.

Started well with few winter losses. Generally wet, but with a freak heat wave in May or June, lasting about a week.

Bees appeared quite strong throughout the summer, but never seemed to store any great quantities of honey at any time. Feeding was required several times during the summer. It would have paid to feed heavily in July. Varroa strips used again, put into hives when ling removed and left for about 5 weeks or longer.

Varroa widespread in Scotland, even in the Islands off the West Coast but none actually seen on our bees.

2002

This was a disastrous season right up until the end of July. Spring and Summer was poor all along, with very low temperatures and a high rainfall. I noted August the 27th as being very hot, a gorgeous day, one of the few until then, with exceptionally heavy rain everywhere and floods in Pitlochry, England, and in Europe, especially Germany.

There was virtually no rape honey at all, and just a small crop of raspberry. For the last few years, the raspberry acreage has been decreasing steadily, and now a lot of what is left is under plastic tunnels, and not so good for bees. Some growers nearby buy bumble-bee nests to use under the plastic.

The honey farm at Cally had to feed bees syrup right through the summer and into August. My own bees did not need a great deal of feeding, just making enough to get by.

Fortunately, August improved, and the ling yielded very well.

2003

Last season at Kirkmichael

Hives wintered very well at Polcalk. No losses, and most strong.

By 20th May most had about 15/20 lbs rape in supers. Split 3 colonies and took nuclei to Ardblair. By June 7th (very early) some had 40 lbs rasp/rape honey.

Moved to a new site at Tarvie (which I had used about 40 years earlier) as the road up Glen Derby was very rough. The weather turned very good and temperature was 80-82 F. (28 C.) in mid-July. This continued on and off until end of August resulting in a wonderful ling crop, all in the comb.

The raspberry crop is now mostly under plastic tunnels, so this is not helping yields, but still plenty of rape around.

Ed. Dad received correspondence from beekeepers from all over the world – from those who had visited the farm, friends he and mum had made on their Apimondia tours and young beekeepers wanting to pick his brains. He heard from Quentin Henderson, who had worked for Honey Farmers and then moved out to the island of Nevis as beekeeping adviser. He was very proud to say that some of the photos he had taken himself had been used on the Nevis stamps and he was actually portrayed on the 80c stamp! He had been very busily occupied over the previous year rebuilding the Beekeepers Headquarters, there which had burnt down.

2004

Moved to Bridge of Cally

This was a very wet summer and extremely disappointing for beekeepers. My own records are poor, as we were busy moving out of "The Sheiling" and back down to Bridge of Cally where I intended to build a new small honey shed.

The heather supers were extracted at Heather Hills Honey Farm.

2005

The Rowans, Bridge of Cally

My 60th year of beekeeping! How time flies – it seems just a few years since my first colony arrived at Hill of Birches in Lanarkshire, by train from the south-west of England.

2005 was a hot dry year with some good weather early on, a not too good May, but a wonderful July and August.

My bees were taken up to the Heather site at Tarvie, to a site I had used about 35 or 40 years earlier. But it turned out to be the wrong decision this time. The heather varied tremendously from place to place, and Tarvie was poor. Most beekeepers did well. Heather Hills Honey Farm produced about 9 tonnes of bell and 18 tonnes of ling honey – 27 in all.

Bees went into winter in good shape, and my new honey shed at Cally was finished, with water and electricity installed. With the new house being erected at Cally, the bees were a bit neglected.

Polcalk – with Ian's 12x10 poly hives in the background.

2006

This year my son, Ian, bought 25 polystyrene hives from Denmark.

He bought 12 Smith colonies from the estate of the late Charlie Macdonald and with the help of bees transferred from some of my best hives, managed to fill all 25 by the heather flow. His bees went to a site at the foot of Mount Blair.

My 9 hives left went back to Tarvie.

This was a wonderful year for bees after July came in – weeks of very hot weather with sufficient moisture to keep the flow going. The best and hottest summer for years.

My own crop would of course have been a lot bigger if bees had not been transferred to the new hives but each of my hives made a wonderful full cut comb super.

2007

Ian increased his number of Danish polystyrene hives up to 60 in time for the heather flow, assisted by some bees transferred from my Langstroth hives. This obviously reduced my own crop by a considerable amount.

Dalmunzie.

The season was good in April, being exceptionally warm – as hot as summer. But May became cold, wet and rather miserable. June was not good, and July even worse. There was no bell or clover crop at all, and bees had to be fed just to keep them alive. I lost my strongest hive even though it was fed some syrup. Surprisingly, queen mating was good.

August started with poor cool weather but changed dramatically in the second half of the month and ling nectar flowed in heavily until about 10th September.

Barbara and I celebrated our Golden Wedding in April.

2008

This was a very poor season and had there not been a good week or two in August it would have been a disaster.

Bees were transferred into Ian's new polystyrene hives to help him get established so this greatly reduced my own early crop.

Weather was poor – cold, wet and windy anyhow so yields were low. There was no raspberry honey, and very little oil seed rape.

Bees had to be fed in summer to keep them alive. Some colonies did not build up well. Queen mating was also poor.

However, there was quite a good crop of cut comb, mostly from about 4 hives only.

2009

Another poor season!

It started well with exceptionally good weather in May and early June, but July was poor, and only the first ten days of August were nice. After that here was a lot of rain and little or no honey was gathered.

We treated for Varroa with Formic Acid, but there was a very severe outbreak of European Foul Brood all over Perth and Angus especially with some further north.

This was treated with Oxytetracycline special licensing by the Government (three applications of the drug mixed with icing sugar).

The early August ling flow was good, and so the cut comb crop was excellent for such a short period.

2010

Quite a good year with a late Spring.

All colonies were treated several times with Formic Acid pads. The EFB of the previous year all seemed to have disappeared, except in three hives which were taken to a "Treatment" apiary at Stylemouth, Ballintuim. There they were shaken onto new foundation and all the brood combs burned.

These colonies made very little honey after that, but they built up good brood combs, having been well fed.

Raspberries did quite well considering the big number of hives in the area of Alyth, and the small acreage outside of the plastic tunnels.

Rape too did quite well.

Hives were moved up to the heather a bit too quickly, which resulted in three queenless hives later. These were just shaken out.

Bell was not too good in this area, but very good in the usual good areas. Ling was reasonable.

2011

This was a most disappointing season.

England had a pretty good year with very high temperatures in late September – up to 29.9 in the south. A record.

Scotland was very different.

There was a very good ten days or so in April when it felt more like mid-summer, but from then on it was very very poor weather, and there were only a few days during the whole summer that one would call summer-like!

The rape yielded pretty well in May, but colonies had been weakened by the European Foul Brood the year before, or the crop would have been bigger. There was almost no raspberry honey due to the wet, cold weather in June, even although there were some fields near our bees without the awful plastic tunnels. It was the poorest raspberry crop ever!

The weather was still terrible in July with heavy rain almost every day! The EFB did not bother Ian and myself very much, but it was still quite widespread in Perth and Angus. There was also some American Foul Brood, but fortunately this did not hit us (yet!)

The heather flow was not too bad in places, but cut comb was hopeless.

2012

Another very poor season.

After the outbreak of European Foul Brood disease in the last two years, there was one this year which had to be destroyed by burning. At the time of burning, it appeared to be recovering well with a very nice new brood nest, but it was destroyed anyhow.

Rape in May/June yielded well, but after that it was dreadful weather with a few nice days here and there.

It was necessary to feed right up until the end of June, as rasps yielded very little. It was wet, wet, wet – the wettest June for years!

Bell heather gave no honey at all, except in Glenesk, and some other places, where some quite good results were obtained.

Varroa was treated several times with a Formic Acid pad, placed above the broodnest – three times over a ten day period.

Trying to produce comb heather was hopeless because of the weather. Only one of the colonies gave about 18 lbs. The other 11 hives packed whatever they got into the broodchambers, which became quite heavy.

Definitely the worst season for years!

Overlooking Strathardle.

2013

Although I did not get delivery of my first colony of bees until VE Day, June 1945, I did start to study bees in 1944, so next year will be my 70th year of beekeeping, I hope!

2013 was a wonderful Summer, but results in honey production did not come up to expectations.

Most colonies were in reasonable strength in Spring, but a few were weak.

Some queens were given away when the colonies got a bit strong, and increase was made.

The problem came, I think, when colonies with a queen cell were moved to the heather. The young queens did not mate well – possibly some were lost, so this much delayed the build up of these parent hives.

Only half the hives were strong enough to make a crop at Pitcarmick.

Rape yielded quite well, raspberries very little.

The weather improved all the time, and up the glen clover yielded very well.

There was not much from the bell heather although it looked good. Other areas did well. So for the first time in many years I got some nice clover honey.

Disease was not a big problem. I simply gave the Formic Acid pads twice during the season. Dosing three times at three-day intervals.

In July the temperature went up to 28/29 degrees for several days.

2014

Ed. By 2014 Dad felt that he was beginning to find the work too heavy, so decided to cut down on the number of his bee hives a little more, though nothing would prevent him from tackling all the other normal tasks that he was accustomed to carrying out around the house and farm.

Results appeared very poor but this is because the 6 strongest colonies were sold in July. The remaining hives had queen cells or young queens and had been split.

It was a very good summer with lots of sunshine and enough rain most of the time.

But July produced no honey for me, although bell and clover yielded well in other areas.

Ian did well with a ton of rape and a ton of mostly ling, all from about 100 hives.

Things to do!

With 3 of my 7 grandchildren – Connor, Rohana and Murray.

2015

70 years of my Honey Record – does not seem possible!

Seems like only a very few years ago that I collected a travelling box at Glasgow Central Station – an 8-frame colony from J.T. Burgess of Exeter. The fact that that was VE Day in 1945 seems now to have been a very good omen.

This was a strange year. Good wintering, with almost no Nosema, which at one time caused big losses. (Nowadays, with good polystyrene insulation under the roof, most losses are due to queenlessness).

The weather in May was quite good, and the bees did very well on the rape. Especially well because they had to build new combs in the supers. (I had decided to sell all my MD depth supers and change over to standard Langstroth depth, the big supers were getting to be too heavy for me in my old age).

However, at the end of May the weather became very unsettled and continued that way well into August. Cold and wet.

Bees had to be fed for a long time or they would have starved.

There was no bell heather honey in Perthshire, but strangely in Glen Esk it gave good yields.

The weather did improve a bit after mid-August, and the ling gave a smaller than average crop. Prices rose accordingly.

2016

This was a very disappointing year.

Bees wintered well and gave a moderate crop of rape considering the poor weather. However, the weather continued to be cold and poor, so that in June when we were expecting some raspberry honey the bees had to be fed nearly all the time, or they would have died. Mating was not good either, so extra queen cells had to be added later.

July was a strange month. Many very heavy downpours which must have washed out the nectar, but good hot days at times.

We obtained no bell honey in our area, but in the Sma' Glen there was a good crop, up by Amulree.

The same sort of weather continued in August. Hot days with very little coming in, and very heavy showers quite frequently.

Some colonies did make a fair bit of ling honey making an average crop but in my case with empty cut comb boxes above, most went into the broodchambers, which became solid.

What little ling that I had was all centrifuged by Ian. I produced no comb honey for the first time in years.

Even the supers are getting a bit too heavy for me now, and I wonder how much longer I can do some actual beekeeping. But Ian always appears at the critical time, so we will see!

2017

This was a very disappointing season.

The rape did well, but there was little or nothing from the few raspberries within reach.

Weather was unsuitable for July, although some beekeepers did get some bell honey.

Poor weather again in August meant a smaller ling crop but the Heather never looked better.

Ed. 2017 marked two major events as he and mum celebrated 60 years of marriage in April and dad reached the grand old age of 90 in October.

Barbara and I celebrated our Diamond Wedding Anniversary in April.

My 90th birthday celebrations in October with friends, Richard and Joyce Wilson, Cych Valley Honey Farm, Wales, as surprise guests.

2018

This was a terrible year for my bees, though the crop in other areas of Scotland varied from poor to very good!

The winter went on a month too long. The farmers were all running out of hay and straw. The bees were nearly all weak but some would have recovered if the weather warmed up and for a few days it was 16-17 degrees. When the colonies should have been expanding, in February we had the "Beast from the East" – cold strong winds and very low temperatures which caused the weaker colonies to die out, and the stronger ones to just survive.

In June I sold 2 of the best colonies, and one empty hive, leaving me with a total of 7 (cutting back on account of age 90!)

The weather in June was poor and feeding had to be carried out from time to time even into July. However, I did have a swarm come into an empty hive! Then, it was not wet, but too dry, and it just got hotter and hotter, with temperatures reaching 30 degrees in June.

There was no nectar in the flowers, and it was the end of the honey-gathering as far as I was concerned. In other parts of the country there was a good flow from bell heather and from ling. Where I had my bees in Strathardle it was hopeless, but even a few miles away on higher hills there was a reasonable flow.

July and August were both much warmer than usual, with no rain. When we had some rain towards the end of August it was too late. The heather never even bloomed...!

Total honey produced
by Athole Kirkwood since 1945 –
1,524,311 lbs
(about 690 tonnes)!

Ed. So, Dad's bee diary ends with an entry describing everything the Scottish weather could possibly throw at a beekeeper, which just goes to show how his chosen life in the glens depended so precariously on a fickle climate.

For the thirty or so years of retirement he enjoyed, until he passed away in March 2020 just before the coronavirus pandemic hit the globe, Dad continued to complete his yearly records, with his operation running at a gradually decreasing scale. The annual Production Records for his entire career can be found in the Appendices at the back of this book. His hard copy will be lodged with the Moir Library at a future date.

Epilogue

2020

It is silent in the honey shed as it has been since the beginning of lockdown and I get a slight start and feel a lump in my throat when the first thing I see is dad's bee suit hanging up to air on the edge of a pile of supers where he last left it. There is a fading aroma of honey, propolis and the unmistakable musty scent of bees hanging on the air and everywhere there are reminders of his particular way of working. The shed is spotless beneath the growing cobwebs – rows of jars containing different types of nails and screws, neatly stacked tins and buckets. Everything in its place. I reflect now that this might be a product of his army days. His gloves and veil are lying on a table, at exact right angles to the edge, his favourite hive tool beside them. I pick it up – I will keep this.

His was a long life, full of hard work and relaxing family time.

He was always willing to try something new – golfing is a sport much loved in the Kirkwood family, (his mother played golf at the very hilly course at Pitlochry in her nineties) but somehow he could never really get the hang of it, a tank driving lesson was much more appreciated and he had enormous fun trundling around the fields. Our favourite memory though is of a white water rafting day out. When we had all changed into our wet suits we turned to see that dad had managed to put his on inside out and was still wearing his brown brogues and flat cap. (He was probably still wearing his tie).

We had many memorable holidays in caravans, cottages and occasionally abroad. As with everything, preparing for any sort of trip was a major operation, with five children packed in the car (a Citroen with three rows of seats!) and all the luggage tied on to the roof in military fashion.

He would every year look forward to attending the bee farming conferences with "Uncle" Norman Stewart from Aberfeldy. On their return, (sometimes bearing small trinkets for us kids) we would all sit at the kitchen table listening to their hilarious tales of their time spent meeting up with their beekeeping chums.

We remember many afternoon teas with mum and dad's beekeeping friends "Uncle" Norman, "Uncle" Andrew Scobbie from Fife, "Uncle" Bert Mason and "Uncle" Bob Cousten, who, a keen member of the Magic Circle, was to us children, the magician at our birthday parties – "Izzy Wizzy Lets Get Busy"! (Sadly, not so many bee Aunties in those days). Richard Wilson, from Wales, a man of many talents, would entertain us with his guitar.

Mum would also perform her role as hostess at Open Days for beekeeping groups and individuals from all over the world – from Sweden to Burma, Tanzania to Trinidad. We were all roped in to make sandwiches and refill endless pots of tea. One of the beekeeper's wives asked if mum ran a hotel – she certainly could easily have done so.

Dad could not have set off to the bees without mum having made his flask of tea and fried egg sandwiches with (dad's favourite) a banana that most would have thought well past its sell by date!

Many travellers dropped in when they saw the sign at the bottom of the road. One American lady enquired if honey was made by boiling up the heather on the moors. Another inadvertent visitor, a European

Athole at Ceanghline

cyclist, who was stung by bees as he cycled up the glen, was nursed back to health by mum after a few days' rest at Glenardle.

Another great beekeeping friend was Francis Cordingley. His son, Marcus, who sadly passed away recently, recalled that Dad told him "You should treat a beehive like a box of eggs" and certainly careful meticulous workmanship and a tidy apiary were insisted upon. His own sons, Ollie and Xavier, are carrying on his legacy in the beautiful village of Scampston, near Malton, where Yorkshire Honey is based.

Marcus, along with an array of other students from all over, stayed during the summer and worked at the bees and in the honey farm. Yukka, a Finnish beekeeper, would take the stairs at breakfast time at one leap, in his clogs, and took an extremely large splinter out of the floorboards. Another "helper" decided it would be a good idea to put a polo mint in the bottom of each jar of honey! If the strange sound of "om" was heard it would be coming from a Hare Krishna devotee to be found in the lotus position in a corner of the honey farm.

There was also, of course, a great deal of work to be done in the office – dad believed in the "flat" system and there was paper everywhere, but all in meticulous order. Mum would deal with the wages but it was also a yearly task to deliver the rent for the apiary sites to the areas around Blairgowrie and to the estate factors in Strathardle, Glenshee and Braemar. Time had to be found for these errands as well. Another aspect to the job was making trips to shops either to deliver their stock, or to drum up new orders. Visits to his agent, Jim Munro, in Glasgow were made periodically. He would take honey to him and the return trip would bring boxes of tinned food of all kinds (very helpful for a large family).

There were of course many misfortunes – bad weather, disease, vandalism, marauding cattle, unpaid bills, pricing difficulties and illness – dad suffered from a painful kidney stone for some time until it was removed (he kept it in a jar and labelled it of course, but I have spared you a photograph of that here), and then a broken foot sustained while fishing at Loch Loch as well. Ian well remembers the day he had to carry dad a considerable distance over the moors on his back after suffering a broken foot.

The most troubling period was when he had thought his busy bee farming days were over and his intention had been to settle down into an early retirement. However, misfortune intervened and decided it had a different future in store for him. Honey Farmers Ltd, who had bought the farm from him, got into difficulties, leaving large debts, and, after eight years, dad decided his best and only option was to buy back what he could, although much was in poor condition and the apiaries had been taken over by others. It was a worrying time.

He spent the next year or two building the farm up again before selling on to my brother Ian in partnership with Bernard and Mark Noonan and retiring again.

In spite of these setbacks, and demoralising circumstances, he never found them insurmountable. His attitude was a pragmatic one and he was always determined to look on the bright side. He could not have succeeded as he did without having an extremely positive outlook on life. Marcus recollected his upbeat attitude from a typical comment – "Well on a day like this there could be a ton of honey coming in!"

He would frequently make us laugh with his anecdotes - one he was fond of telling was from a time when the men were hammering in posts around an apiary to keep the cattle out and a passerby commented "You'll never keep them in with that"!

Dad often said that he still felt the same in himself as he always had done, and until very late in life, he still possessed remarkable strength of mind and body. There were few tasks in the farm or at home that he would not take on. It is of course hard to read in his later diary entries that he was beginning to find the work too heavy. There is a regret in his words which predicts a time when he would be unable to properly tend his beloved bees. However, thankfully, that fate never befell dad as he passed away towards the end of winter after a relatively short illness and his bees are now tended by his son.

Dad, though of course a product of his time, always kept abreast of the latest beekeeping literature and was keenly interested in the most modern developments. He recognised that bee husbandry was of necessity continually evolving. The techniques he used himself worked for the environment in which he found himself at that time.

Although, in the years during his building up of his business, the development of chemicals, plastics, large scale farming and all the commodities which that and future generations have felt they could not do without, came on in leaps and bounds, beekeepers were of course among the first to notice that not all progress was beneficial and that sprays, polytunnels and the disappearance of wild areas in our countryside, would prove costly. He would frequently describe to us the animals and birds which used to be common in our locality

Bluebell Wood, Muirton

and the changes, always for the worse, in the landscape. Noticeably, as the years went by, the troubled shakes of the head after returning from the berries, in particular, would become more and more frequent.

It is certain though, that he never had any regrets as to his chosen path in life. He was exactly where he wanted to be with the wife he adored and the children and grandchildren he did so much for. When he was awake, he was constantly busy, performing tasks around the house or at the bees. The only exceptions to this would be either entertaining visitors or watching Countdown or the Antiques Roadshow, always with a cat on his lap. Otherwise, he was renowned for snoring in his chair, again with a cat on his lap.

Despite all the warnings dad received that full time beekeeping north of the border would be "impossible", these only spurred him on to prove his critics wrong.

He was single-mindedly determined to make a good living from his vocation and did so. The ancestral traits he inherited from his forebears of an excellent mathematical mind, along with engineering, analytical and financial skills enabled him to grow the business in a logical and thrifty manner. His relationship with the land and his deep awareness of his own history in the glens supplied the unswerving passion he felt for his lifelong devotion to his bees.

In Memory of beekeeping friends lost to us since Dad's passing
Bob Barrie
Richard Wilson
Brian Sherriff
Quentin Henderson, The Beeman of Nevis
Marcus Cordingley
and of course childhood pal, Eddie Millington

With love from Barbara,
Fiona, Ian, Catriona, Sarah and Valerie,
grandchildren Craig, Rohana, Murray, Connor,
Lauren, Holly and Erin,
and great grandchildren, Taylor and Finlay

With thanks to Willie Robson for his thoughtful obituary
and for his help in identifying some of the beekeepers and honey
farms in the previous photographs

APPENDIX I

Production Records 1945-2018

Production figures refer to actual honey produced. They have little to do with sales. They do include any seed added but do not include any honey purchased eleswhere. *Weight in imperial measure changing to metric over time.

Year	1945	1946	1947	1948	1949
Location	Hill of Birches		Vale of Leven		Albert Drive/ Loch Thom
Winter Hive Losses					
Spring Hive Count	1	1	2	2	3
Smith					
Langstroth					
Bought					
Sold					
Spring Count Total	1	1	2	2	3
Smith					
Smith nuclei					
9-frame Dadant					
Langstroth					
Autumn Count Total	1	2	2	3	3
Average Yield (lbs) *Spring Count*	35	20	15	10	55
Average Yield *Previous Autumn Count*				10	55
Clover Honey	35	20	30	21	48
Clover & Raspberry					
Raspberry Honey					
Rasp/Oil Seed Rape					
Bell Heather Honey					
Ling Heather Honey					
Heather Honey Total					108
Sugar fed in summer					
Swarms					
TOTAL CROP*	35	20	30	21	156
Total honey since 1945					

Year	1950	1951	1952	1953	1954
Location	Bridge of Weir		Bridge of Weir		Bridge of Weir/ Blairgowrie
Winter Hive Losses					
Spring Hive Count	3	2	26	28	30
Smith					
Langstroth					
Bought		20			50
Sold					
Spring Count Total	3	22	26	28	80
Smith					
Smith nuclei					
9-frame Dadant					
Langstroth					
Autumn Count Total	2	32	40	37	100
Average Yield (lbs) *Spring Count*	33	14	22	27	7
Average Yield *Previous Autumn Count*	33	14	17	19	6
Clover Honey	93	10	128	0	232
Clover & Raspberry					
Raspberry Honey					
Rasp/Oil Seed Rape					
Bell Heather Honey					
Ling Heather Honey					
Heather Honey Total	6	18	414	760	236
Sugar fed in summer					
Swarms					
TOTAL CROP*	99	28	560	768	468
Total honey since 1945					

Year	1955	1956	1957	1958	1959
Location	Blairgowrie	Blairgowrie	Blairgowrie	Blairgowrie	Blairgowrie
Winter Hive Losses					
Spring Hive Count	4	190	215	266	386
Smith					
Langstroth					
Bought		130	32	25	
Sold					
Spring Count Total	134	222	240	424	386
Smith					
Smith nuclei					
9-frame Dadant					
Langstroth					
Autumn Count Total	216	273	286	520	520
Average Yield (lbs) Spring Count	100	26	58	26	100
Average Yield Previous Autumn Count	50	22	47	25	80
Clover Honey	3249	2616	5047	4032	14500
Clover & Raspberry					
Raspberry Honey					
Rasp/Oil Seed Rape					
Bell Heather Honey				4032	4950
Ling Heather Honey				2604	1600
Heather Honey (All)	4641	2256	8300		
Comb Honey	690	805	1084	420	3250
Sugar fed in summer					
Swarms					
TOTAL CROP*	3 Tons 17 Cwts	2 Tons 11 Cwts	6 Tons 9 Cwts	4 Tons 19 Cwts	17 Tons 5 Cwts 60 lbs
Total honey since 1945					

Year	1960	1961	1962	1963	1964
Location		Achalader			
Winter Hive Losses		7%	15%	31%	11.5%
Spring Hive Count	567	640	727	694	920
Smith					
Langstroth					
Bought			43		
Sold					
Spring Count Total	567	640	770	694	920
Smith					
Smith nuclei					
9-frame Dadant					
Langstroth					
Autumn Count Total	670	850	1009	1040	1123
Average Yield (lbs) *Spring Count*	74	25	43	34	63
Average Yield *Previous Autumn Count*	62	24	37	23	56
Clover Honey	23184	4144	8064	14884	16464
Clover & Raspberry					
Raspberry Honey					
Rasp/Oil Seed Rape					
Bell Heather Honey	9520	0	20160	500	23408
Ling Heather Honey	8025	9372	1344	5000	17584
Heather Honey (All)	4641	2256	8300		
Comb Honey	1025	44	300		1380
Sugar fed in summer				6 Tons	
Swarms					
TOTAL CROP*	18 Tons 12 Cwts 90 lbs	6 Tons 1 Cwt 48 lbs	13 Tons 4 Cwts	9 Tons 2 Cwts	26 Tons 5 Cwts 36 lbs
Total honey since 1945		135971	165539	186923	244759

Year	1965	1966	1967	1968	1969
Location		Achalader	Bridge of Cally		
Winter Hive Losses	14.5%	15%	4.8%	11.5%	8%
Spring Hive Count	968	950	1183	1212	1235
Smith		1075	1046	1046	1046
Langstroth				20	300
Bought					
Sold					
Spring Count Total	968	950	1183	1212	1235
Smith		1075	1046	1046	1046
Smith nuclei			131	96	259
9-frame Dadant		170	194	179	
Langstroth				20	300
Autumn Count Total	1115	1243	1371	1341	1605
Average Yield (lbs) *Spring Count*	40	63	66	56	52
Average Yield *Previous Autumn Count*	34	54	62	49	48
Clover Honey	10640				
Clover & Raspberry		8290	5264	11760	26432
Raspberry Honey					
Rasp/Oil Seed Rape					
Bell Heather Honey	13664	11760	20832	14784	6384
Ling Heather Honey	13664	40320	50400	40596	31136
Heather Honey (All)	4641				
Comb Honey	380	220	1008	90	90
Sugar fed in summer	5 Tons	7 Tons 15 Cwts	3 Tons	2 Tons	4 Tons
Swarms					
TOTAL CROP*	17 Tons 2 Cwts	27 Tons 1 Cwt	34 Tons 12 Cwts	30 Tons 30 lbs	28 Tons 11 Cwts 90 lbs
Total honey since 1945	283107	343697	421201	488431	552473

Year	1970	1971	1972	1973	1974
Location	Bridge of Cally		Bridge of Cally		
Winter Hive Losses	18%	17%	Langs 8% Smth 20%	Langs 10% Smth 28%	15% mainly Smith
Spring Hive Count	1110				
Smith	1046	838	808	512	301
Langstroth	500	442	692	919	1329
Bought					
Sold					140
Spring Count Total	1110	1280	1500	1431	1630
Smith	1046	1043	990	672	193
Smith nuclei					
Langstroth nuclei					
Langstroth	500	750	1020	1271	1484
Autumn Count Total	1546	1793	2010	1943	1677
Average Yield (lbs) *Spring Count*	50	75	48	59	27
Average Yield *Previous Autumn Count*	41	62	41	48	25
Clover Honey					
Clover & Raspberry	14336	13946	8304	12824	17646
Raspberry Honey					
Rasp/Oil Seed Rape					
Bell Heather Honey		15740	8332	15064	10416
Ling Heather Honey	40320	65970	55706	56000	57000
Heather Honey (All)					
Comb Honey	220	275	120	380	
Sugar fed in summer	6 Tons 10 Cwt				
Swarms					
TOTAL CROP*	24 Tons 14 Cwt	42 Tons 16 Cwt 59 lbs	32 Tons 7 Cwt	38 Tons 5 Cwt 44 lbs	20 Tons 2 Cwt 38 lbs
Total honey since 1945	607349	703280	775742	860010	905072

Year	1975	1976	1977	1978	1979
Location			Bridge of Cally		
Winter Hive Losses	24%	24%	16.4%	28%	43%
Spring Hive Count					
Smith					
Langstroth	1300		1718		
Bought		70 (Norman)			275
Sold					200
Spring Count Total	1300	1640	1718	1310	1140
Smith					
Smith nuclei					
Langstroth nuclei					
Langstroth					
Autumn Count Total	2025	2050	2016	1877	1830
Average Yield (lbs) *Spring Count*	98	48	40	49.7	35
Average Yield *Previous Autumn Count*	76	38	33	34.7	21
Clover Honey					
Clover & Raspberry	15570				
Raspberry Honey		9352	17646	13840	16960
Rasp/Oil Seed Rape					
Bell Heather Honey	36624			15224	
Ling Heather Honey	74816	67950	50110	35900	22840
Heather Honey (All)					
Comb Honey	672	700	660	160	
Sugar fed in summer					
Swarms					
TOTAL CROP*	57 Tons	34 Tons 16 Cwt 49 lbs	31.033 Tonnes	29.54 Tonnes	18.05 Tonnes
Total honey since 1945	1,032,754	1,110,756	534.87	564.41	582.46

Year	**1980**	**1981**	**1982**	**1983**	**1984**
Location			Bridge of Cally		
Winter Hive Losses	33%	20%	23%	16%	6%
Spring Hive Count					
Smith					
Langstroth	850	126	119	130	154
Bought					
Sold	370				
Spring Count Total	850	126	119	130	154
Smith					
Smith nuclei					
Langstroth nuclei				7	9
Langstroth					
Autumn Count Total	1445	161	154	154	154
Average Yield (lbs) *Spring Count*	29.1	80	67	75	166
Average Yield *Previous Autumn Count*	19.3	60	52	63	166
Clover Honey					
Clover & Raspberry					
Raspberry Honey	3730	1800		4900	
Rasp/Oil Seed Rape			3312		8000
Bell Heather Honey	6830	1580	3150	1272	6480
Ling Heather Honey	14160	6070	1558	3600	11060
Heather Honey (All)					
Comb Honey				28	32
Sugar fed in summer	7 Tons			6 lbs each	
Swarms					
TOTAL CROP*	11.21 Tonnes	4.59 Tonnes	3.64 Tonnes	4.445 Tonnes	11.584 Tonnes
Total honey since 1945	593.68	598.26	601.9	606.34	617.924

Year	1985	1986	1987	1988	1989
Location	Balnald	The Sheiling			Bridge of Cally
Winter Hive Losses	5%	28%	25%	7%	13%
Spring Hive Count					
Smith					
Langstroth					
Bought				100	30
Sold					
Spring Count Total	154	115	120	250	450
Smith					
Smith nuclei					
Langstroth nuclei	5		10		
Langstroth					
Autumn Count Total	155	160	160	483	730
Average Yield (lbs) *Spring Count*	65	25	76.8	77.8	66
Average Yield *Previous Autumn Count*	65	19	57.6	70	61
Clover Honey					
Clover & Raspberry					
Raspberry Honey		1344			
Rasp/Oil Seed Rape	9000		3840	14960	14850
Bell Heather Honey	460	582	2203		6555
Ling Heather Honey	600	1040	3150	4500	7920
Heather Honey (All)					
Comb Honey			24		150
Sugar fed in summer					
Swarms					
TOTAL CROP*	4.564 Tonnes	1.345 Tonnes	4.283 Tonnes	8.829 Tonnes	13.37 Tonnes
Total honey since 1945	622.675	624.020	628.202	637.032	650.405

Year	1990	1991	1992	1993	1994
Location		Kirkmichael			
Winter Hive Losses	5.5%	10%	11%	10%	6%
Spring Hive Count					
Smith					
Langstroth					
Bought					
Sold					
Spring Count Total	690	31	30	30	32
Smith					
Smith nuclei					
Langstroth nuclei			3		1
Langstroth					
Autumn Count Total	762	34	34	34	34
Average Yield (lbs) *Spring Count*	89.8	86.7	70	81	93
Average Yield *Previous Autumn Count*	84.9	79	61	72	87
Clover Honey					
Clover & Raspberry					
Raspberry Honey			764		
Rasp/Oil Seed Rape	30300	800	560	1480	1320
Bell Heather Honey	14500	837	350	550	1010
Ling Heather Honey	17000	240	257	200	118
Heather Honey (All)					
Comb Honey	200	800	158	216	528
Sugar fed in summer					
Swarms					
TOTAL CROP*	28.13 Tonnes	1.22 Tonnes	0.947 Tonnes	1.109 Tonnes	1.35 Tonnes
Total honey since 1945	678.536	679.675	680.97	681.63	682.96

Year	1995	1996	1997	1998	1999
Location					
Winter Hive Losses	6%	15%	8%	0%	0%
Spring Hive Count					
Smith					
Langstroth					
Bought					
Sold					
Spring Count Total	32	29	34	34	24
Smith					
Smith nuclei					
Langstroth nuclei					3
Langstroth					
Autumn Count Total	34	34	23	24	24
Average Yield (lbs) *Spring Count*	118	33	127	68	39
Average Yield *Previous Autumn Count*	111	28	117	68	39
Clover Honey					
Clover & Raspberry					
Raspberry Honey					
Rasp/Oil Seed Rape	2240	860	2026	1610	560
Bell Heather Honey	727		350		370
Ling Heather Honey	355	30	120	12	10
Heather Honey (All)					
Comb Honey	476	70	555	22	
Sugar fed in summer					
Swarms					
TOTAL CROP*	1.71 Tonnes	0.435 Tonnes	1.384 Tonnes	0.746 Tonnes	0.426 Tonnes
Total honey since 1945	684.69	685.31	686.51	687.44	687.87

Year	2000	2001	2002	2003	2004
Location					Bridge of Cally
Winter Hive Losses	0	0	0	0	0
Spring Hive Count					
Smith					
Langstroth					
Bought					
Sold	12				
Spring Count Total	24	12	12	12	12
Smith					
Smith nuclei					
Langstroth nuclei	2		2		
Langstroth					
Autumn Count Total	12	12	12	11	12
Average Yield (lbs) *Spring Count*	20	49	28	110	25.1
Average Yield *Previous Autumn Count*	20	49	28	110	25
Clover Honey					
Clover & Raspberry					
Raspberry Honey					
Rasp/Oil Seed Rape	60	427	74	929	112
Bell Heather Honey	120	116		157	20
Ling Heather Honey	14	14	102	108	40
Heather Honey (All)					
Comb Honey	40	46	160	340	30
Sugar fed in summer					
Swarms					
TOTAL CROP*	0.106 Tonnes	0.269 Tonnes	0.152 Tonnes	0.696 Tonnes	0.137 Tonnes
Total honey since 1945	687.98	688.249	688.401	688.91	689.047

Year	2005	2006	2007	2008	2009
Location					
Winter Hive Losses	0	0	0	0	0
Spring Hive Count					
Smith					
Langstroth					
Bought					
Sold					
Spring Count Total	12	12	11	10	10
Smith					
Smith nuclei					
Langstroth nuclei					
Langstroth					
Autumn Count Total	11	9	10	9	13
Average Yield (lbs) *Spring Count*	47	80	50	22	34
Average Yield *Previous Autumn Count*	47			25	27
Clover Honey					
Clover & Raspberry					
Raspberry Honey					
Rasp/Oil Seed Rape	260	140	130	70	120
Bell Heather Honey	100	80		20	28
Ling Heather Honey	114	56	100	16	24
Heather Honey (All)					
Comb Honey	102	220	217	122	175
Sugar fed in summer					
Swarms					
TOTAL CROP*	0.256 Tonnes	0.225 Tonnes	0.391 Tonnes	228 lbs	347 lbs
Total honey since 1945	689.303	689.528	689.919	690.022	690.18

Year	2010	2011	2012	2013	2014
Location					
Winter Hive Losses	0	0	0	0	0
Spring Hive Count					
Smith					
Langstroth					
Bought					
Sold					
Spring Count Total	11	9	12	11	13
Smith					
Smith nuclei					
Langstroth nuclei					
Langstroth					
Autumn Count Total	10	12	12	12	7
Average Yield (lbs) *Spring Count*	88	9	35	49	35
Average Yield *Previous Autumn Count*	75	8	35	45	30
Clover Honey					
Clover & Raspberry				310	
Raspberry Honey					
Rasp/Oil Seed Rape	640	60	405	160	180
Bell Heather Honey	100				
Ling Heather Honey	60	20	8	12	22
Heather Honey (All)					
Comb Honey	175		10	58	24
Sugar fed in summer					
Swarms					
TOTAL CROP*	975 lbs	80 lbs	423 lbs	540 lbs	220 lbs
Total honey since 1945	1,522,132	1,522,212	1,522,635	1,523,175	1,523,395

Year	2015	2016	2017	2018	
Location					
Winter Hive Losses	0	0	0	0	
Spring Hive Count					
Smith					
Langstroth					
Bought					
Sold				3	
Spring Count Total	6	6	7	7	
Smith					
Smith nuclei					
Langstroth nuclei					
Langstroth					
Autumn Count Total	8	8	10	4	
Average Yield (lbs) *Spring Count*	71	36	30	10	
Average Yield *Previous Autumn Count*	53	27	21	6	
Clover Honey					
Clover & Raspberry					
Raspberry Honey					
Rasp/Oil Seed Rape	346	144	150	60	
Bell Heather Honey		24			
Ling Heather Honey	65	52	35		
Heather Honey (All)					
Comb Honey	15		25		
Sugar fed in summer					
Swarms					
TOTAL CROP*	426 lbs	220 lbs	210 lbs	60 lbs	
Total honey since 1945	1,523,821	1,524,041	1,524,251	1,524,311	

APPENDIX 2

William W Smith

An excerpt from *Reflections on Beekeeping*

by kind permission of W S Robson

Willie Smith was Scotland's first commercial beekeeper and the designer of the Smith hive. His friend, William Selby Robson, felt he should have received more credit for what he achieved. He was a dispatch rider in the first war in France and thereafter he became a chauffeur to a mill owner in Innerleithen who encouraged Willie to keep bees during the daytime when he wasn't needed as a driver. He wished to be a progressive beekeeper and designed a new hive which was made according to British Standard dimensions but to the American pattern. The hive was very cheap to make being four pieces of wood only. He went for top bee space as this prevented the box above coming into contact with the lugs of the frames below allowing the box to be tipped up and inspections made for queen cells; thus the frames in the bottom box were not dislodged by the upward force and no bees were crushed. The only drawbacks of the Smith hive are the poor handholds, and a deep box weighs 70lb when filled. His hives were made of close-grained white pine and painted with white lead paint as in the American tradition. In the spring Willie put on a brood chamber of frames from which the heather honey had been scraped. Peebles, where Willie had his bees, is a late area and this provided a huge boost for the bees. Readers may know of the American tradition to provide pollen patties in the spring. Willie's method amounted to the same thing. Thereafter, the queen had three boxes to work in, the bottom box being largely brood-free thus providing ventilation and some considerable control of swarming. Normally, the queens were replaced by the supersedure impulse but if they did build cells he would find the queen and make an artificial swarm.

Sometimes the queens built cells in July out of frustration due to unsettled weather. This provided him with huge problems as he needed to keep the bees together. It meant going through every hive and shaking the bees off the combs and removing the cells until the bees lost interests in swarming. By this time the hives were very powerful and Willie only had a net veil and bare hands. This is unimaginably difficult work. He was a very big strong man and worked 120 hives single handed. In Peebleshire, before massive afforestation, the bell heather came nearly to the valley bottom with the ling on the tops. Permanent sites therefore made a great deal of sense. The honey was pressed, his target being 5 cwt. a day. Thereafter, the jars of heather honey were put into warm water baths and sold as clear heather honey with air bells. The grocer in Innerleithen dressed the complete shop window with his honey. This was often photographed.

Willie's bees would be selected from the local strain and yet R O B Manley wrote to my father saying he had never encountered such powerful colonies. This spoke volumes for Willie Smith's ability and the plentiful flora in that locality at that time. The object was to have massive colonies ready for the heather. The great honey farms in the south of England would keep mainly imported bees in large Dadant hives yet their colonies were not as strong as Smith's and his bees were of a local indigenous strain entirely acclimatized to Peebleshire, a late and difficult area for beekeeping. Willie Smith kept 150 hives of fees reducing to 120 hives nearing the end of forty-year commercial beekeeping career.

See *Reflections on Beekeeping* by W S Robson, Northern Bee Books, ISBN 978-1-904846-82-6 for more detailed material and reminiscences regarding William W Smith.

Obituary

Athole M. C. Kirkwood

I was sorry to hear of the death of Athole Kirkwood. Right back into the 1950s I remember my parents talking about his progress in the world of commercial beekeeping. Heather Hills Honey Farm was a revelation in Scotland at that time – the first attempt to build a big honey farm in the North, in the most beautiful county of Perthshire.

No doubt he would buy some local colonies to get going (subclinical AFB was common in Perthshire at that time) and finding them unsatisfactory would buy French Blacks from Steele and Brodie in Dundee. They were the main hub in beekeeping at that time and the inevitable winter losses in the East of Scotland were made up by French Blacks. Later on he obtained some Caucasian queens from Mountain Grey Apiaries and settled on these crosses.

At the start of the season all the queens' wings were clipped (2,200 colonies) to prevent them swarming, prior to making splits. Winter would hang on in Perthshire and this would be a stressful time. Generally, the bees would rear their own queen cells but some queens (virgins?) were obtained from Wilson Carne, a queen breeder in Cornwall. As time went on colonies were united and taken to the bell heather in early July. There would be intermittent honey flows along the way but none that would give a crop, apart from the raspberries. All the effort went into getting heather honey. I remember him telling me that he and his men had taken 2,000 hives to the heather using VW pickups. A Herculean task by any standards, no hive lifters or 4WD. In most years there was a crop, in some years a bonanza and in others the bees had to be fed throughout the season.

Meanwhile, a new purpose-built honey house was constructed with all new machinery to extract heather honey. I remember the beautifully engineered Lifetime tangential extractors that allowed the combs to be reversed when the extractor was going at full speed! When there was a big crop of honey, he found that he was unable to extract the heather honey in good time before it crystallized in the comb. Sometimes extracting went on into November by which time it was too late. So began the huge task of putting the bees into Langstroth hives and converting all the machinery as well as automating the perfor extractor so the extracting could finish on time using very many fewer combs for the same volume of honey. Athole started out using Smith hives, the industry standard at the time. I think all the hives he ever had were factory made using first grade cedar (Thorne). They will still be in use somewhere.

Throughout this period of dynamic expansion Athole was supported by his wife Barbara and a wonderful gentleman called Norman Stewart from Aberfeldy who knew the business inside out. I shouldn't forget the foreman either, Bob I remember. During the early 1980s a firm of investors called Honey Farmers Ltd were looking to buy beekeeping businesses and Athole took the fateful decision to sell up, no doubt because of the level of responsibility and variable seasons. Honey Farmers started off well as 1984 was a wonderful season but the following 3 seasons were impossibly difficult and Heather Hills, with Steele and Brodie, drifted into administration.

Athole and his son Ian bought most of the business back and soldiered on. Athole would be disappointed. At Chain Bridge we felt for him, as one does. However the business continues today under new ownership and Ian has a small honey farm up the road at Kirkmichael.

Athole Kirkwood was an exceptionally successful beekeeper to build such a prestigious business in a difficult part of the country. He was a giant among beekeepers.

W. S. Robson
Chain Bridge Honey Farm

Bibliography

Australasian Beekeeper, The (15 October 1975) 'Apimondia Congress – Grenoble'. *Australasian Beekeeper,* Vol. 77, No. 4, p. 75.

Bill, Lesley E. (1989). *For the Love of Bees, the Story of Brother Adam of Buckfast Abbey.* David and Charles Publishers PLC. ISBN 0-7153-9261-1.

Brother Adam. (1969). *Meine Betriebsweise.* Franz Ehrenwirt Verlag.

Brother Adam. (1975). *Beeping at Buckfast Abbey.* British Bee Publications. ASIN B0007AK3PU.

Brother Adam. (1983). *In Search of the Best Strains of Bees, and the results of the evaluations of the crosses and races.* Northern Bee Books. ISBN 0-907908-06-3.

Brother Adam. (1987). *Breeding the Honeybee.* Northern Bee Books. ISBN 0-907908-32-2.

Clements, D.A. (2013). *A Man and his Bees – The Story of Alec Wilfred Gale.* Northern Bee Books.

Digges, J.G. Manley, R.O.B. (1949). *The Practical Bee Guide – a Manual of Modern Beekeeping.* Talbot Press; fifteenth edition (1949). ASIN B001NGWVMA.

Field, Oliver S. Field. (2022). *Notes on Queen Rearing.* Ibra & Nibb. ISBN-10 1913811131.

Furzey, John. *Beekeeping for Beginners and Old Hands* (DVD). Countryside Audio & Visual. ASIN B0042TPK32.

Henderson, I.A.N. (October 1976) 'Honey from the Hills'. *The Scots Magazine,* Vol. 106. No. 1. pp. 61-69.

Kirkwood, A.M.C. (1969-1970) *Honey Looseners.* Lincolnshire Beekeepers' Association Yearbook, p. 35.

Kirkwood, A.M.C. (January 1959) *Swarm Control on a Scottish Bee Farm.* The Glasgow Beekeepers' Review.

Kirkwood, A.M.C., NDA NDD (1986) *The Hive of the Future?* The 1986 Beekeepers Annual, ISBN O-907908-34-9, ISBN 0265-8690, pp. 95-99.

Kirkwood, A.M.C. (12 January 1952) 'Honey Market Report - Scottish Counties'. *The British Bee Journal,* No. 3627, Vol. LXXX, p. 25.

Kirkwood, A.M.C. (11 December 1952) 'Honey Market Report – Scottish Counties'. *The British Bee Journal,* No. 3675, Vol. LXXX, p. 793.

Kirkwood, A.M.C. (January 1986) 'What Other People Think'. *The Scottish Beekeeper*, Vol. LXIII, p. 7.

Kirkwood, A.M.C. (1973) "A Commercial Beekeeping Enterprise in Scotland" in *Bee World* Vol 54 Issue 3.

Louveaux, J., Maurizio, A. and Kirkwood, A.M.C. (1973), 'Heather: Bees-Honey-Beekeeping. Bee Research Association reprint'. *Bee Research Association.* Bee World, 54(3) : 105-120 (1973).

Manley, R.O.B. (1936). *Honey Farming in the British Isles.* Bradley and Sons. ASIN B00087UMBW.

Manley, R.O.B. (1946). *Honey Farming.* Faber & Faber Ltd. ASIN B0016JL8Z0.

Manley, R.O.B. (1948). *Bee-Keeping in Britain.* Faber & Faber Ltd. ASIN B000XZWEDI.

Noonan, M. & Kirkwood, A.M.C. (August 2019), 'Heather Hills Farm'. Interviewed by Marshall, G. with Ellis, A., *Bee Farmer,* Vol. 5, Issue 4. pp. 5-9.

Rice, Norman V., MBE. *Queens' Land – Aussie Bee Beaudesert.* Northern Bee Books. ISBN 978-0-907908-68-5. 2013.

Robson, W.S. (2011). *Reflections on Beekeeping.* Northern Bee Books. ISBN 10 1904846823/ ISBN 13 9781904846826.

Robson, W.S., (2022). *His Words.* Northern Bee Books. ISBN 978-1-914934-48-3.

Ruttner, Friedrich, Milner, Eric & Dews, John E. (1990) *The Dark European Honey Bee.* The British Isles Bee Breeders Association, ISBN 0-905369-08-4, p. 4.

Skilling, Robert N.H., FRSA, FSCT (ed.) (March 1978) 'Can Scotland Support Bee Farming?' *Scottish Bee Journal,* p.4.

Skilling, Robert N.H., FRSA, FSCT (ed.) (March 1978) 'A Man and his Bees'. *Scottish Bee Journal,* p. 5.

Tonsley, Cecil C. (ed.) (17 May 1969) 'Our Cover Picture'. *British Bee Journal.* No. 4191, Vol. XCVII.

Tonsley, Cecil. (2020). *Honey for Health.* Northern Bee Books. ISBN 978-1-912271-67-2.

Willis, Gavin. *People and Families of Mattishall.* https://mattishall-village.co.uk

Newspaper and Journal Articles

Ariel (1972) 'A Close Run Thing with 100 Million Bees!' *Ariel,* 8 December

Beck, George (1971) 'Britain's Biggest Honey Producer aims for 2000 Hives'. *The Scottish Farmer,* 25 December, p.22.

Blairgowrie Advertiser (1962) 'Scotland's Largest Honey Farm Moving to New Site'. *Blairgowrie/ Perthshire Advertiser,* 6 April.

Blairgowrie Advertiser (1969) 'Blairgowrie Honey Farm has 1,400 Hives – Talk to Rotarians'. *Blairgowrie Advertiser,*7 March.

Courier and Advertiser Supplement (1968) 'Taking the Lid off a Honey Farm'. *Dundee Courier,* 13 March, p. 3.

Dundas, Peter (1959) 'Does a beehive hold the secret of long life?' *Scottish Sunday Express*, Glasgow, 16 August.

Evening Telegraph (1962) 'Bees by the Million!' *Evening Telegraph*, 24 July, p. 10.

Farming Independent (1977) 'A1 for Apple Blossom'. *Farming Independent*, No. 4, February.

Farming Leader (1971) 'Bees in Blairgowrie – A Visit to Athole Kirkwood'. *Farming Leader*, June, pp. 36-38.

Fiona (1971) 'Speaking Personally by Fiona'. *Dundee Courier*.

Food Trade Review (May 1968) 'Success of Honey Farm in Scotland', *Food Trade Review.*

Gaisford, Michael (1977) 'More than one Profit from a Honey Harvest', *Farmers Weekly*, April 1977, pp. (viii)-(xi).

Mercer, David (1968) 'Harvest of the Hives – Perthshire Honey Farm's Success'. *The Scotsman*, 23 February.

Nature Man, The (1975) 'The Holy Terror of Glen Tilt'. *The Sunday Post*, 5 October, p.11.

People's Journal (1969) 'You Won't See any Animals on this Bridge of Cally Farm'. *The Peoples Journal*, 14 June, p.11.

Perthshire Advertiser (1979) 'Beeline to Perth for Honey'. *Perthshire Advertiser*, 16 February

Robertson, Bill (1976) 'Kirkwood's Kingdom – 100 Million Workers Spell Sweet Success'. *Daily Record*, 15 July, p. 16.

Scots Magazine (1976) 'Honey from the Hills', I.A.N. Henderson, *Scots Magazine*, October Issue

Sunday Post (1962) 'Now Mr Kirkwood has the Biggest Bzzzness in Scotland'. *The Sunday Post*, 24 June.

Sunday Post (1965) 'Glasgow Boy Turns £7 10s into a £25,000 Business – Now he Employs 100 Million Workers!', *Sunday Post*, 9 May, p. 11.

Sunday Post (1972) 'HON Makes Friends with the Queen! He Even Helps her Spring-Clean.' *The Sunday Post*, 23 April.

Times (1968) 'Exports from Scottish Bees by a Scottish Correspondent'. *The Times*, 25 March.

Weekly News (1982) 'Joanna Buzzes off for a Day with the Bees'. *The Weekly News*, 21 August, p. 13.

Weekly Scotsman, The Scotsman (1961) 'His Workers are Swarming to Comb up the Market'. *The Scotsman*, 27 July.

Acknowledgements

Archive Department
D C Thomson & Co Ltd
80 Kingsway East
Dundee DD4 8SL

Scottish Beekeepers' Association

Excerpts from the letters of R. O. B. Manley by kind permission of Morgan Manley
Excerpts from the letters of A. W. Gale by kind permission of
Peggy Carter and Mary Gale
Excerpts from the letters of E. W. D. Madoc by kind permission of Sally Beevor

W. S. Robson, Chain Bridge Honey Farm

Scotland's Rural College (SRUC) (formerly Auchincruive College)
The King's Buildings
West Mains Road
Edinburgh EH9 3JG

John Swire & Sons Ltd
and Special Collections, University of Bristol Library (www.hpcbristol.net).

David Moore
Paul Metcalf

www.ingramcontent.com/pod-product-compliance
Ingram Content Group UK Ltd.
Pitfield, Milton Keynes, MK11 3LW, UK
UKHW022031041026
15591UKWH00012B/629

9 781918 815016